DATE DUE

Y0-CXG-284

NEW METHODS IN
ANALYTICAL CHEMISTRY

NEW METHODS IN
ANALYTICAL
CHEMISTRY

RONALD BELCHER

B.Sc., Ph.D., F.R.I.C., F.Inst.F.

Reader in Analytical Chemistry,
University of Birmingham

AND

CECIL L. WILSON

Ph.D., D.Sc., F.R.I.C., F.I.C.I.

Reader in Analytical Chemistry,
The Queen's University of Belfast

London : Chapman & Hall Ltd.
REINHOLD PUBLISHING CORPORATION
NEW YORK

First Published 1955
Second Printing, 1956

PRINTED IN GREAT BRITAIN
BY JARROLD AND SONS LIMITED, NORWICH

Reprinted in U.S.A.
Edwards Brothers, Inc.
Ann Arbor, Michigan

PREFACE

BOTH the scope of analytical chemistry and its literature have increased at such a rate recently that it is very difficult for text-books of analytical chemistry to keep pace with the new developments. Most of the standard textbooks may be placed in one of two broad classes. There are, in the first class, those which deal with the newer modes of analysis, usually referred to under the inclusive heading of instrumental or physical methods. Such books, if they are to act as working manuals, must of necessity confine themselves to one or two modes or instruments, but a good range of specialised works on such analytical fields as absorptiometry, spectrographic methods, polarography or chromatography is available.

The second category comprises the conventional textbooks of analytical chemistry. Unless devoted to one field of analysis, such as the analysis of metals and alloys, or the analysis of fuels, these books are particularly serviceable for teaching purposes. But their range as working manuals is often restricted by the fact that they tend to confine themselves primarily to methods of long standing, only replacing these by more modern methods when the latter have been thoroughly tested under a wide variety of conditions. It is possibly for this reason that the average analytical chemist tends to show surprise when he is told that there is much that is new in the fields of gravimetric and titrimetric analysis; there tends to be a belief that all that is new in analytical chemistry must needs function round an instrument.

Consequently a new method employing the classical techniques may remain hidden in the literature until some research worker is faced with the problem of determining a particular constituent under conditions hitherto regarded as unusual, and where the established methods fail. The worker resurrects the method and may find that it meets his requirements. Such resurrection frequently necessitates many hours of searching the literature, in addition to the experimental work involved.

Both of us have on many occasions found it necessary to seek alternative methods for a variety of determinations, and the initial

v

task has not been made lighter by finding, from time to time, that the journal sought is not readily accessible, or is even "missing from the shelves". It seemed to us, therefore, that there was a distinct place for some compilation of the newer methods based on classical techniques. The idea is not an original one. In 1932 a book with a similar aim by A. D. Mitchell and A. M. Ward covered the methods developed since the First World War, and earned the gratitude of analytical chemists. The warm reception which this excellent book received encouraged us to attempt a comparable task.

It must be stressed that we use the term "new methods" in a very flexible sense. The obvious point of time from which to start is the date of publication of the book by Mitchell and Ward. And this has, in general, set our limit, although a high proportion of the methods which we describe are, in fact, of quite recent origin. Our main concern, however, has been to include many methods which have not yet reached the standard textbooks, and which are therefore not generally accessible.

No claim is made that we have tested every method described and have found its claims justified. To undertake systematically such a range of testing would afford a lifetime's work. On the other hand we have had occasion to use a reasonable number of the methods, and as a result of our findings certain methods which were not entirely satisfactory—these, happily, were few—have been omitted from our final list.

We must hasten to add, however, that the omission of any particular new method does not necessarily mean that we have found it unserviceable. We may inadvertently have missed it. Or it may have occurred in some field of analytical chemistry outside our usual run of interests. In addition, some process of selection had perforce to be applied in order to keep the size of the book within reasonable bounds. In deciding which methods to include, among those of which we had no direct experience, we have naturally been guided partly by personal preference, since one is always attracted to some fields more than others. Within this range some discrimination has also been used. We have had no hesitation, for example, in including such methods when recommended by investigators of repute.

It is our hope that by making the working details of these methods readily accessible in book form we may enable many more analysts to test them for their own purposes than would otherwise be possible. That there is a great demand for information concerning new methods, and that a remarkable number of analytical chemists find difficulty in keeping abreast of the literature, are supported by the earlier history of a small portion of the material included in this book. During the past five years several lectures were given by one of us in which some of these methods were described. These lectures were given in some dozen centres in Britain (generally under the auspices of the Royal Institute of Chemistry) as well as at a week-end course at the Royal Military College of Science. The attendance and interest in every case were gratifying. In addition, when a report of one lecture was published (*Journal of the Royal Institute of Chemistry*, 1950, **74**, 139) more than fifty letters were received asking for references to the methods.

This information has always been freely given, and it is understood that in consequence, for example, some of the material described in that lecture has already appeared in a new edition of at least one standard textbook published in Great Britain. We feel that it is proper to mention this, lest our claim to cover non-textbook material may appear unfounded, or at least exaggerated. Up to the present, as far as discrimination has been permitted to us, most of the methods which we describe have not yet been considered suitable for inclusion in the standard textbooks on quantitative inorganic analysis.

As far as possible, methods have been presented in a uniform fashion, in order to make for ease of working in preparing solutions and in carrying out experimental directions. The uniform mode of presentation which has been aimed at is to refer first to previous methods where this is relevant, and to give the important properties of the reagent to be described. The compound formed or the reaction involved is included where known, together with any special properties which may have a bearing on its analytical behaviour. The optimum conditions are next discussed, together with the reasons for choosing these, if known. Ions which interfere and those which do not inter-fere, and the occurrence and avoidance of unfavourable conditions are detailed, and finally the treatment of the product is dealt with. This overall discussion is followed by the detailed experimental procedure recommended, and finally if an unusual reagent is proposed its synthesis is included where possible. Literature references are listed sectionally for ease of reference.

Certain sections, such as that on indicators, do not readily lend themselves to precise treatment according to this scheme. But even apart from these exceptions, it will readily be observed that this uniform method of presentation has not been achieved. The divergent usages employed by different authors, and, indeed, frequently appearing within a single paper, prevented any near approach to ideal description. Thus in some cases interferences are listed and in others no indication has been given as to whether these have been investigated. Some authors present a clear and reasoned statement of the background of the analytical procedure; others do not. In some cases normalities of acid concentration, in others simple content by volume of "concentrated" acid, and in yet others pH have been quoted. Again, even within a single paper all of these usages may be found.

Where it seemed to us that variants of the method of noting quantities or concentrations, or of stating other details, could be brought into a reasonable uniformity without serious risk of making a vital error in procedure, this has been done. All too often, however, it has been necessary to leave the details precisely as described in the original paper, or to omit what seemed to us desirable information because it was not included by the authors of the method.

In a very few instances, indeed, where what seemed to us an important method might have been omitted because experimental details were either wanting or were described in confused fashion, we have even ventured to deduce the missing steps and include them here or to edit the description drastically.

As a result of all this, a considerable and frequently an undesirable heterogeneity of treatment may be observed throughout these pages. We hope that study of these collected methods may, by comparison, encourage authors of analytical papers to give some thought to the mode of presentation of a new analytical method so as to include all the necessary information for others who may wish to use it.

We also hope to be informed of any errors which may have crept in through our transliteration of original papers. And we will welcome any comments by our readers on the satisfactory or unsatisfactory operation of any of these methods.

We would express our thanks to those authors who have kindly sent us reprints of their publications; and in particular to Dr. A. J. Nutten and Dr. T. S. West, who have permitted us to make extensive use of their reviews of certain sections of analytical chemistry; and to the latter, Dr. D. Gibbons, and Dr. W. W. Harpur, who read the proofs and made many helpful observations on these.

R. B.
C. L. W.

CONTENTS

ix

CHAPTER I

SEPARATION BY PRECIPITATION

PRECIPITATION in analytical procedures may be carried out for one of a variety of purposes, of which the most important are: (1) to produce a suitable weighing form; (2) to separate an ion in a form which can then be further treated for determination either gravimetrically or titrimetrically; or (3) to remove an interfering ion prior to the determination of another ion. Precipitation methods which belong essentially to the first of these categories will generally be found dealt with elsewhere in this book.

Of methods falling into the second and third categories, interest may centre either in the use of a novel precipitant or in the use of a novel method of precipitation. In the present chapter, procedures will be confined to those in which one or other of these interests may be regarded as inherent, regardless of whether the precipitate produced is the final analytical form or not.

Of novel methods of precipitation, that which is undoubtedly of most interest at the present time is so-called "controlled precipitation" or "precipitation in homogeneous solution".[1] When a precipitant is added in the ordinary course of an analytical procedure, local concentrations at the point of addition produce abnormal conditions of precipitation which frequently result in co-precipitation through adsorption or occlusion. In some cases serious effects may be avoided by the frequently advised slow addition of reagent and constant stirring while addition is taking place. Certain precipitates, however, cannot be produced without suffering seriously from contamination, even under the most favourable experimental conditions, and with the greatest precautions.

In precipitation in homogeneous solution either the reagent which is added is not the precipitating agent, but is converted to this after addition, more or less slowly, by an intervening reaction; or the desirable conditions for precipitation, such as pH, are gradually produced after adding the precipitating reagent by the controlled

I

reaction of another substance. In this way the precipitating agent is evenly distributed through the solution before precipitation takes place. There is the added advantage that the concentration of reagent actually or effectively increases gradually from zero to the amount required for complete precipitation, this increase also being evenly distributed throughout the solution.

Such precipitations are becoming increasingly used, and are obviously an important new form of attack on the sources of error in a fundamental analytical operation.

[1] H. H. Willard, *Analyt. Chem.*, 1950, **22**, 1372: L. Gordon, *ibid.*, 1953, **24**, 459.

1. PRECIPITATION OF *m*-NITROBENZOATES

m-Nitrobenzoic acid

was one of the first organic compounds to be proposed for the precipitation of an inorganic cation, having been recommended by Neish many years ago for the selective precipitation of thorium in the presence of lanthanons.[1] Until recently it has been little used.

Recently Osborn[2] has investigated the reagent as a precipitant for quadrivalent elements in the presence of lanthanons and other tervalent cations and of bivalent cations. It has been shown that under suitable conditions cerium-IV, thorium and zirconium are quantitatively precipitated in hot solution. Preliminary examination suggests that hafnium is also precipitated. No precipitate is formed with a number of tervalent lanthanons examined, or with aluminium, antimony, beryllium, bismuth, cadmium, cobalt, lead, lithium, magnesium, manganese, palladium, potassium, thallium-I, titanium-III, titanium-IV, uranium or yttrium.

Hydrolysis of the salts, and not the action of the cation, causes precipitates to form with tin-II and tin-IV. Mercury, either in the mercurous or the mercuric form, precipitates partly in the cold. This precipitate dissolves on boiling, but reappears on cooling, and the precipitation is then quantitative. It is therefore possible to determine this element in the presence of other elements precipitated by the reagent.

In each of the precipitations investigated the acid concentration is very critical. By making use of the solubilities in different acid concentrations, thorium, which will not precipitate quantitatively above 0·02 N and is not precipitated at all in 0·1 N acid, can be separated from zirconium, which will precipitate quantitatively in 0·2 N acid, and is 90 per cent precipitated in N acid. Use may also be made of the fact that cerium-III is not precipitated while cerium-IV is quantitatively precipitated, to separate this element from thorium or zirconium.

Plutonium-IV behaves like the other quadrivalent elements, being precipitated on zirconium-IV carrier, and precipitation is over 95 per cent complete at pH 1·88.[3] Plutonium-III is not precipitated.

It is of interest to note that salicylic acid, like *m*-nitrobenzoic acid, gives precipitates with cerium-IV, hafnium, thorium and zirconium, but not with cerium-III.[4]

Reagents

Reagent solution. Dissolve 4 g. of *m*-nitrobenzoic acid in 1 litre of water, heat to 80° C., allow to stand overnight, and filter.

Wash solution. Dilute 50 ml. of reagent solution to 1 litre with water.

Procedures

Precipitation of thorium. Dissolve the sample, containing about 0·5 g. of thorium, in 100 ml. 0·02 N nitric acid, or preferably in water. If the sample is already in solution, adjust the volume to 100 ml. and the acidity to 0·02 N or less. Add 150 ml. reagent solution, heat to 80° C., set aside for 1 hour, and filter. Wash the precipitate with 100 ml. wash solution, place the wet filter paper and precipitate directly in a platinum crucible, dry, ignite, and cool. Weigh as ThO_2.

$$mg. \ ppt. \times 0·8789 \equiv mg. \ Th.$$

Precipitation of zirconium. Dissolve the sample in 100 ml. water or nitric acid up to 0·2 N. Carry out the precipitation as for thorium.

Filter through a Whatman No. 44 filter paper, wash and ignite. Weigh as ZrO_2.

$$\text{mg. ppt.} \times 0.7403 \equiv \text{mg. Zr.}$$

Precipitation of cerium. Convert the cerium to the quadrivalent form if necessary, and determine in the same way as for thorium. Weigh as CeO_2.

$$\text{mg. ppt.} \times 0.8142 \equiv \text{mg. Ce.}$$

Precipitation of mercury. Dissolve the sample in 100 ml. 0.02 N nitric acid or adjust the acidity to this value or less. Add 150 ml. reagent solution in the cold, boil, and allow to cool to room temperature. Allow some time for the precipitate to re-form, filter and wash with cold wash solution. Redissolve the precipitate and determine the mercury by any standard method.

Separation of thorium and zirconium. Dissolve the sample in 100 ml. 0.2 N nitric acid and add 150 ml. reagent solution. Heat to 80° C., set aside for 1 hour, and filter on a Whatman No. 44 filter paper. Wash, ignite, and weigh the precipitate as ZrO_2. Add excess ammonia to the filtrate; filter, wash, ignite and weigh the precipitate as ThO_2.

Separation of cerium and thorium or zirconium. Reduce the cerium to the tervalent form, and precipitate the other element. Reoxidise the filtrate to cerium-IV, and precipitate this by the normal procedure.

Separation of mercury from thorium, zirconium or cerium. Precipitate in the normal manner, and heat to dissolve the mercury compound. Filter while hot, and treat the precipitate for the appropriate element. Allow the filtrate to cool, filter off the mercury compound, and treat as already described.

[1] A. C. Neish, *J. Amer. Chem. Soc.*, 1904, **26**, 780.
[2] G. H. Osborn, *Analyst*, 1948, **73**, 381.
[3] B. G. Harvey, H. G. Heal, A. G. Maddock and E. L. Rowley, *J.C.S.*, 1947, 1010.
[4] A. Jewsbury and G. H. Osborn, *Anal. Chim. Acta*, 1949, **3**, 642.

2. PRECIPITATION BY CONCENTRATED NITRIC ACID

Considerable difficulty has always been experienced in the separation of strontium and calcium, and some form of extraction has usually been employed, the most widely known being the extraction of the nitrates by a mixed ether-ethanol solvent (p. 35).

Willard and Goodspeed[1] have found that strontium may be separated quantitatively from calcium and a wide range of other cations by precipitation with concentrated nitric acid. Barium and lead are also quantitatively precipitated. For complete precipitation of strontium the optimum concentration of nitric acid in the solution is 80 per cent. Below 79 per cent the solubility of the strontium nitrate is appreciable, and above 81 per cent the solubility of calcium nitrate is too small. Barium is best precipitated at a nitric acid concentration of 76 per cent, and lead at 84 per cent. The method may be used for any one of these three in the absence of the other two.

It is necessary to add the concentrated nitric acid drop by drop with constant stirring until the acid concentration is brought up to the required value.

Where 60 mg. of strontium are present, a suitable final volume is about 40 ml., and in these conditions one obtains efficient separation from 25 mg. aluminium, 75 mg. tellurium, 100 mg. chromium, 150 mg. lithium or silver, 250 mg. copper, 300 mg. sodium, and at least 500 mg. ammonium, antimony, arsenic, beryllium, bismuth, cadmium, cerium-IV, cobalt, iron-III, lanthanum, magnesium, manganese, mercury, nickel, potassium, selenium, thallium-III, tin-IV, uranium or zinc. If more than 25 mg. calcium is present a double precipitation is necessary. Separation from titanium is not satisfactory. Interferences in the determination of barium and lead have not been fully investigated, but are presumably comparable. Chloride produces low results in the determination of lead, and must be absent, so that the separation of lead from antimony or tin is not feasible.

Procedures

Separation of strontium. Evaporate the chlorides, perchlorates or preferably the nitrates to dryness in a 50-ml. beaker, and dissolve in 10 ml. water. Precipitate strontium nitrate by slow dropwise addition of 26·0 ml. concentrated nitric acid, with vigorous and constant stirring. Set aside for half an hour, or, if the precipitate is very small, for 45 minutes with continued stirring throughout the first 15 minutes. Filter through a Gooch crucible, transferring the precipitate by washing with a jet of 80 per cent nitric acid. Wash with ten 1-ml. portions of this wash liquid, dry at 130°–140° C. for 2 hours, and weigh as $Sr(NO_3)_2$.

$$mg.\ ppt. \times 0.4140 \equiv mg.\ Sr.$$

Separation of barium. Dissolve the dry salts in 5 ml. water. Add slowly, with constant stirring, 3·0 ml. 70 per cent nitric acid, followed by 11·0 ml. concentrated acid, thus bringing the overall acid concentration to 76 per cent. Set aside for half an hour, filter, and wash the precipitate with ten 1-ml. portions of 76 per cent nitric acid. Dry for 2 hours at 130°–140° C., and weigh as $Ba(NO_3)_2$.

$$mg.\ ppt. \times 0.5255 \equiv mg.\ Ba.$$

Separation of lead. Dissolve the dry salts in 2·5 ml. water. Add 5·0 ml. 70 per cent nitric acid drop by drop, followed by 13·0 ml. concentrated nitric acid, bringing the overall acid concentration to 84 per cent. Allow to stand for half an hour, filter, and wash the precipitate with ten 1-ml. portions of 84 per cent nitric acid. Dry for 2 hours at 135° C., and weigh as $Pb(NO_3)_2$.

$$mg.\ ppt. \times 0.6256 \equiv mg.\ Pb.$$

[1] H. H. Willard and E. W. Goodspeed, *Ind. Eng. Chem. Anal.*, 1936, **8**, 414.

3. CONTROLLED HYDROLYTIC PRECIPITATION OF HYDRATED OXIDES

When a boiling solution containing platinum, palladium, rhodium and iridium as chlorides, together with sodium bromate, is adjusted to approximately pH 7·0, the hydrated oxides of the last three metals precipitate quantitatively, leaving the bulk of the platinum in solution.[1] A second precipitation makes the separation from platinum complete. Precipitation of rhodium and iridium is actually complete at pH 6·0, but this is a borderline value for palladium, which is completely precipitated at pH 8·0.

This method gives a much better separation of platinum than the usual ammonium chloroplatinate method in which the precipitate is liable to considerable contamination and is appreciably soluble. The method may be included in a schematic analysis of the six platinum metals, and its accuracy is comparable with that of the best analytical procedures for commoner metals.

In such a schematic procedure osmium is first removed as the tetroxide by distillation from nitric acid solution, followed by removal of ruthenium, also as the tetroxide, by distillation from a solution in dilute sulphuric acid containing sodium bromate. Both distillates are absorbed in 6 N hydrochloric acid saturated with sulphur dioxide, and are subsequently precipitated as hydrated oxides and ignited to the metals for determination. Palladium, rhodium and iridium are then separated from platinum by the controlled hydrolytic precipitation method. From the solution of the hydrated oxides palladium is precipitated as the dimethylglyoxime complex, and either weighed as such or converted to the metal. Rhodium is precipitated as sulphide and ignited to the metal, or is reduced to the metal by titanous chloride in dilute boiling sulphuric acid solution. Iridium is reprecipitated as hydrated oxide and ignited to the metal. From the residual solution platinum is precipitated as the sulphide and is ignited to the metal.

Procedure

Separation of palladium, rhodium and iridium from platinum. Evaporate the solution to a moist residue on the steam bath. If nitric acid is present evaporate twice with 5 ml. concentrated hydrochloric acid to remove all nitrogen. Add 2 g. sodium chloride and 5 ml. concentrated hydrochloric acid, and dilute with water to 300 ml. Heat to boiling. Add 20 ml. filtered 10 per cent sodium bromate solution. Add carefully 10 per cent sodium bicarbonate solution until the dark green solution shows evidence of the formation of a permanent precipitate. Test the acidity of the hot solution from time to time by spotting with bromocresol purple solution, and stop the addition of bicarbonate when the indicator colour changes from yellow to blue (pH 6·0). Add a further 10 ml. bromate solution and boil for 5 minutes. Add bicarbonate solution dropwise until the solution just shows a faint pink with cresol red (pH 8·0). Boil for 15 minutes. (If palladium is absent the complete precipitation may be carried out at pH 6·0.)

Allow the precipitate to settle for a few minutes, and filter through a porous-porcelain filter. Wash the precipitate with hot 1 per cent sodium chloride solution adjusted to pH 6·0–7·0.

Redissolve the precipitate in the original beaker by heating with 10–20 ml. concentrated hydrochloric acid on the steam-bath. Transfer the crucible to another beaker and leach out twice by heating on the steam-bath with 5 ml. concentrated hydrochloric acid. Combine the leachings with the main bulk of solution, add 2 g. sodium chloride, and evaporate to dryness on the steam-bath. Add 2 ml. hydrochloric acid, dilute to 300 ml. with water, and repeat the precipitation of the hydrated oxides.

Add 20 ml. hydrochloric acid to the filtrate from each precipitation and warm carefully. After effervescence has subsided, concentrate the filtrates somewhat and combine them. Evaporate to dryness on the steam-bath, and repeat the evaporation with hydrochloric acid to destroy all bromate. Dilute the solution somewhat and filter, washing the filter with 0·1 N hydrochloric acid. Dilute to 400 ml. Add 20 ml. hydrochloric acid and precipitate the platinum as sulphide.

[1] R. Gilchrist and E. Wichers, *J. Amer. Chem. Soc.*, 1935, **57**, 2565.

4. PRECIPITATION OF BASIC FORMATES

The precipitation of the basic formates has been used for the separation of both bismuth and iron-III.

Most methods for the separation of bismuth are based on the hydrolysis of bismuth salts in weak nitric acid solution, and such methods are of limited use, particularly in the presence of lead. Prior separation of the lead by precipitation as sulphate is liable to considerable error. A separation based on the precipitation of basic bismuth acetate was proposed by Herzog[1] but was not very satisfactory. The precipitation of bismuth as basic formate was very much superior[2] and has recently been re-examined and modified.[3] In .the modified method the acid solution is neutralised with ammonium hydroxide and ammonium carbonate. Ammonium formate is added to precipitate the bismuth, and the basic formate can then either be ignited to the oxide, or redissolved for determination as oxychloride or by some other standard method. Lead can subsequently be determined in the filtrate by precipitation as chromate.

Small amounts of bismuth can be separated from 10–15 g. lead, though in extreme cases considerable care must be given to the neutralisation process.

Ferric basic formate can be precipitated from homogeneous solution as a dense precipitate which is easy to wash and filter, and which does not readily adsorb impurities.[4] This behaviour can be used as the basis for the removal of iron prior to the determination of a number of bivalent metals, although it is not recommended for the determination of iron itself.

The method consists in adjusting the pH of the solution, containing formic acid, ammonium chloride and urea, by means of hydrochloric acid, to a value just below that at which the basic ferric formate begins to precipitate. This ranges from pH 2·0 for an initial concentration of 0·22 mg. iron per 400 ml. to pH 1·55 for an initial concentration of 0·66 g. iron per 400 ml. On boiling, the urea is gradually hydrolysed, and the pH increases steadily. The amount of added urea is calculated to provide a final pH of 4·0. At pH 3·0–3·2 the precipitation of iron is substantially complete. Ten minutes before the final pH is reached, hydrogen peroxide is added to oxidise the small amount of iron which may have been reduced by chloride ion, except in the presence of manganese, where this would oxidise manganese-II and increase loss by adsorption.

Separations of 0·2 g. iron from 1 g. barium, calcium or magnesium, 0·1 g. iron from 1 g. cadmium, cobalt, manganese, nickel or zinc, and 0·1 g. iron from 0·1 g. copper are possible using this method.

Separation from copper is complicated by the formation, about pH 4·0, of a basic copper chloride, which prevents the method from being applied when the concentration of copper is high.

Separation from chromium-III or from chromium-VI (produced by peroxide) is not as efficient as the other separations; appreciable amounts of both these forms are adsorbed.

The amounts of ions adsorbed increase as the pH and the amount of precipitated iron increase. Because of this, a fractional or "two-stage" precipitation has been devised for very precise work, though it is not generally necessary. In this procedure, the pH is allowed to rise to 2·5, and the bulk of the precipitate is then filtered off. The pH is further increased to precipitate the small amount of iron still remaining in solution. Adsorption is very much reduced, since there is little precipitate present at the higher pH range where such loss might be anticipated.

Thorium may be precipitated as the basic formate by urea hydrolysis,[5] to give a dense precipitate, easy to handle, and more satisfactory than that given by the acetate, propionate or succinate. The precipitate begins to form at pH 4·45–4·50, and precipitation is complete at pH 5·3. The method has been recommended for the separation of thorium from the lanthanons after precipitation of these elements by oxalate hydrolysis (p. 13). Cerium-IV must be reduced to cerium-III, hydriodic acid being recommended for this purpose. As a single precipitation does not give a complete separation of thorium from cerium, a double precipitation should be carried out.

The basic formate method is not wholly satisfactory for thorium. The precipitate appears to have a "scavenging" effect on the walls of glass vessels. Consequently silica must be volatilised before final weighing of the thorium oxide. For this reason, quantitative separation of thorium from the lanthanons is more conveniently carried out using tetrachlorophthalic acid (p. 28).

Reagent solutions

Ammonium formate solution. Add 100 ml. water to 300 ml. 90 per cent formic acid solution. Neutralise to litmus with 12 N ammonia (about 650 ml.).

Urea solution. 10 per cent aqueous, freshly prepared.

Procedures

Separation of bismuth from lead. Add 4 N ammonia solution to the warm nitric acid solution of bismuth (containing up to 400 mg.) till a permanent precipitate forms. Just redissolve this precipitate with 8 N nitric acid, added drop by drop. Add 5 per cent ammonium carbonate solution till a further drop would cause a permanent precipitate to form. Boil for 5 minutes to drive off carbon dioxide, disregarding any precipitate which may form. Add 7–8 ml. ammonium formate solution, heat to boiling, and allow to stand on a boiling water-bath for 15 minutes. If very small amounts of bismuth are present, extend this period of heating to 2 hours. Filter through a Whatman No. 42 filter paper, and wash five times with hot water.

Redissolve the precipitate in the original beaker, using 5–6 ml. 8 N nitric acid.

Repeat the precipitation of the basic formate, combining the two filtrates. Wash the precipitate eight times with hot water and once with ethanol. Burn off the filter paper cautiously in a tared porcelain crucible, moisten the residue with 8 N nitric acid, dry on a hot plate and ignite. Weigh as Bi_2O_3.

$$mg. \ ppt. \times 0 \cdot 8970 \equiv mg. \ Bi.$$

To the combined filtrates and washings add 30 ml. concentrated nitric acid and evaporate until heavy white fumes appear. Add 5 ml. 16 N nitric acid and 30 ml. 33 per cent ammonium acetate solution. Dilute to 250 ml., boil, and precipitate lead by addition of excess dichromate solution. Boil for 2–3 minutes till the precipitate turns orange, filter on a tared Gooch crucible, wash with hot water, and dry at 105°C. Weigh as $PbCrO_4$.

$$mg. \ ppt. \times 0 \cdot 6411 \equiv mg. \ Pb.$$

Separation of iron. To the solution add 2 ml. formic acid (sp. gr. 1·20), 10 g. ammonium chloride and the amount of urea indicated by Table I. Follow this by sufficient hydrochloric acid just to prevent formation of the ferric basic formate by adjusting the pH to the desired value. Heat to boiling. If optimum conditions have been achieved, a turbidity should appear 5–10 minutes after boiling has commenced. Transfer the beaker to an air bath which encloses most of the portion of the vessel holding the liquid. Place a stirring rod with an indentation at the end (made by a 1-mm. carpet tack) in the beaker, to prevent bumping of the solution, and allow steady boiling to continue for from 2 to $2\frac{1}{4}$ hours after the first appearance of turbidity. Ten minutes before the end of this time add 5 ml. 3 per cent hydrogen peroxide solution (except where separating from manganese). Filter the precipitate on a coarse filter paper, and wash fifteen times with hot 1 per cent ammonium nitrate solution (adjusted to pH 4·0). With large amounts of precipitate add a little paper pulp just before filtering.

Determine the metals in the filtrate by standard procedures.

Two-stage precipitation of iron. Prepare the solution as before, but with the amount of urea indicated by Table II, Column 2. Boil for the appropriate time, and filter off the precipitate containing the bulk of the iron. Add the second portion of urea solution indicated by the table, and continue boiling to give a total boiling time of $2\frac{1}{4}$ hours. Filter off the remainder of the iron precipitate.

TABLE I

Fe, g.	Urea, 10 per cent soln., ml.
0·11	40
0·33	55
0·66	75

TABLE II

Fe, g.	Urea, 10 per cent solution, ml. First addition	First boiling time, mins.	Urea, 10 per cent solution, ml. Second addition	Total boiling time, mins.
0·11	25	50	15	135
0·33	35	100	20	135
0·66	50	110	25	135

[1] M. Herzog, *Chem. News*, 1888, **58**, 129.
[2] A. L. Benkert and E. F. Smith, *J. Amer. Chem. Soc.*, 1896, **18**, 1055.
[3] S. Kallmann, *Ind. Eng. Chem. Anal.*, 1941, **13**, 897.
[4] H. H. Willard and J. L. Sheldon, *Analyt. Chem.*, 1950, **22**, 1162.
[5] H. H. Willard and L. Gordon, *ibid.*, 1948, **20**, 165.

5. PRECIPITATION OF OXALATES

Oxalic acid $(COOH)_2$, was first recommended for the precipitation of copper as oxalate many years ago,[1] but the method has not been widely applied. Edwards and Gailer[2] have re-examined this precipitation with a view to using it for the removal of copper before estimating tin in brasses and bronzes. Such a precipitation would have the additional advantage that tin forms a stable soluble complex with the reagent in dilute acid solution.

A very small residue of copper is not removed, but its effect is nullified by the addition of ammonium thiocyanate solution immediately before determining tin. The determination is carried out titrimetrically using standard iodine or iodate after reduction of the tin with a nickel spiral.

The method has also been shown to be suitable for the removal of copper from brasses and bronzes before the determination of aluminium, iron and manganese. None of these elements is precipitated under the conditions described, and they may subsequently be determined in the filtrate by standard analytical procedures, after destruction of the excess oxalic acid. One sample suffices for the determination of aluminium, iron, manganese and tin. Nickel and zinc are partly precipitated along with the copper.

Methyl oxalate has been recommended[3] for the precipitation in homogeneous solution of thorium and the lanthanons from a monazite sand extract, before separation of thorium from the lanthanons by a basic formate (p. 10) or a tetrachlorophthalate (p. 28) separation. The method first proposed was subsequently modified[4] somewhat to ensure complete precipitation of all the lanthanons, and thus of any thorium which might otherwise be retained and post-precipitated. At pH 0·8 there is no post-precipitation of lanthanons.

The oxalates are precipitated from a perchloric-hydrochloric acid solution. Since in the first precipitation some phosphate is occluded, a double precipitation is necessary. The precipitate obtained is much more granular and easy to handle than the normal precipitate obtained by the use of oxalic acid itself.

Methyl oxalate appears to be generally applicable to the precipitation in homogeneous solution of elements which give insoluble oxalates. It does not, however, give a satisfactory separation of magnesium from sodium and potassium, apparently because its decomposition is too rapid. Ethyl oxalate, because of its greater stability, seemed likely to be more suitable for this purpose. The method of Elving and Caley[5] has been adapted by Gordon and Caley[6] to take advantage of the fact that ethyl oxalate will precipitate magnesium quantitatively from 85 per cent acetic acid solution. The high concentration of acetic acid is necessary for the decomposition of the ethyl oxalate. The precipitate, which is dense and coarsely crystalline, is easily filtered and washed. The magnesium may subsequently be determined by ignition to the oxide, or preferably by solution of the precipitate in dilute sulphuric acid and titration of the liberated oxalic acid.

Ammonium acetate must be added to adjust the pH and to prevent solution of the magnesium oxalate in a solution of such high acid concentration. Sulphate interferes with the method.

Calcium may be precipitated in homogeneous solution as oxalate by the hydrolysis of methyl oxalate.[7] Well-formed crystals are produced which can be readily filtered and washed, and 20 to 100 mg. calcium can be separated in a single precipitation from 100 mg. magnesium. The precipitation is effected at an initial pH of 4·7 in an acetic acid-ammonium acetate buffer. An excess of methyl oxalate is used to obtain quantitative precipitation and to complex magnesium in order to prevent its post-precipitation.

Reagent

Methyl oxalate, if partly hydrolysed, should be recrystallised from methanol and stored in a dry atmosphere.

Procedures

Separation of copper. Dissolve 1 g. of sample in the form of fine drillings, by warming with 15 ml. 50 per cent sulphuric acid and 10–15 ml. 100-volume hydrogen peroxide in a 400-ml. conical beaker. If the sample is a brass, do not allow the temperature to exceed 40° C. during this operation.

When solution is complete evaporate to small bulk to remove excess of peroxide. Dilute with 10 ml. water and add 10 ml. 50 per cent hydrochloric acid. Add 100–150 ml. 10 per cent oxalic acid solution, and boil gently for 15 minutes. Cool quickly, make up to 250 ml. with water, and allow to settle. Filter through a Whatman No. 40 filter paper. Reject the first few ml. of filtrate, and use aliquots of the filtrate for subsequent determinations.

Breakdown of Monazite Sand. Weigh 1 g. of finely powdered (100-mesh) monazite sand into a 250-ml. conical flask. Add 5 ml. concentrated nitric acid and 30 ml. 70 per cent perchloric acid. Bring gently to the boil, and continue boiling for 1 to 1½ hours after the appearance of dense white fumes. Cool, add 15 ml. water and 5 ml. 3 per cent hydrogen peroxide. Warm gently, and if most of the solid matter does not go into solution, add 5–10 ml. concentrated hydrochloric acid. Cool, and filter to remove zirconia, silica and titania. Wash the precipitate thoroughly with cool 1–3 per cent nitric acid.

Precipitation of thorium and the lanthanons. Evaporate the filtrate from the monazite sand breakdown to white fumes in a 600-ml. beaker. Cool, and add 50 ml. water followed by 5 ml. 3 per cent hydrogen peroxide. Warm gently for several minutes till the solution becomes clear, and add 100 ml. water. Cool and add concentrated ammonia dropwise till a small amount of gelatinous precipitate appears and does not redissolve. Add rapidly 10 ml. concentrated hydrochloric acid, and set aside for 5 minutes.

Add 6 g. methyl oxalate. Stir continuously, and warm gently to 70°–85° C., continuing the heating for half an hour after the first appearance of a precipitate. Add a hot solution of 8 g. oxalic acid in 200 ml. water. Stir, and keep warm for half an hour. Cool to room temperature, adjust the pH to 0·8–0·9 by dropwise addition of dilute ammonia solution, and filter through an ashless filter paper. Wash ten times by decantation with cool 2 per cent oxalic acid solution adjusted with hydrochloric acid to pH 1·0.

Transfer the filter paper with any precipitate back to the beaker containing the bulk of the precipitate. Add 20 ml. concentrated nitric acid followed by 5 ml. 70 per cent perchloric acid. Again warm until fumes of perchloric acid are freely given off, and cool. Carry out the precipitation of the oxalates as before, and wash the precipitate five times by decantation. Transfer the precipitate and filter paper to the beaker once more, and add 20 ml. concentrated nitric acid and 10 ml. 70 per cent perchloric acid. Evaporate till white fumes of perchloric acid are given off, to decompose the filter paper and oxalates. Use this solution for subsequent separation of thorium from the lanthanons.

Separation of magnesium. Separate calcium as oxalate by the usual analytical procedure. Remove ammonium salts and oxalate completely by treatment with nitric and perchloric acids, and take down to dryness. Dissolve the residue of magnesium and alkali perchlorates in 14–15 ml. water. Add 85 ml. glacial acetic acid containing 1 g. dissolved ammonium acetate. Follow this by 1 ml. pure diethyl oxalate (free from ethyl hydrogen oxalate, which is formed by contact of the reagent with moist air, and which decomposes too rapidly).

Stir thoroughly, cover the beaker, and heat to 100° C. Maintain at this temperature for 2 hours after the commencement of precipitation if less than 50 mg. magnesium are present, and for 3 hours if the magnesium content is 100 mg. Fifteen minutes before the end of the heating add 5 ml. 85 per cent acetic acid saturated with ammonium oxalate at room temperature, to ensure complete precipitation.

Filter the hot solution through an X3 sintered-glass filter, and wash four or five times with warm (70°–80° C.) 5-ml. portions of 85 per cent acetic acid. Complete the determination of magnesium gravimetrically as MgO, or titrimetrically. In the latter case dissolve the precipitate in 200 ml. warm (80° C.) 5 per cent sulphuric acid, and titrate immediately with permanganate that has been standardised against sodium oxalate at 70°–80° C.

$$\text{mg. ppt.} \times 0.6032 \equiv \text{mg. Mg.}$$
$$1 \text{ ml. N } KMnO_4 \equiv 12.16 \text{ mg. Mg.}$$

Separation of calcium. Dilute the solution containing calcium and magnesium as chlorides to 150 ml., and adjust the pH to 4·7 with dilute ammonia solution or hydrochloric acid. Add 100 ml. acetic acid-ammonium acetate buffer, 2·5 M with respect to each. Then add 10 g. pure methyl oxalate. Cover the beaker and heat on a regulated hot plate for $2\frac{1}{2}$ hours at 90° C. with occasional stirring. As a precautionary measure, in case the required temperature has not been maintained, add 10 ml. 5 per cent ammonium oxalate solution 10 minutes before filtration.

Cool the solution rapidly to room temperature. Filter through a weighed porous-porcelain filter of medium porosity, and wash with 1 per cent ammonium oxalate solution. Dry the precipitate for 1 hour at 110° C., then ignite at 500° C. for 2 hours. Cool the crucible, moisten the precipitate with several drops of saturated ammonium carbonate solution, heat in an oven at 110° C. for 1 hour, and weigh as $CaCO_3$.

$$\text{mg. ppt.} \times 0.4004 \equiv \text{mg. Ca.}$$

[1] G. Bornemann, *Chem.-Ztg.*, 1899, **23**, 565.
[2] F. H. Edwards and J. W. Gailer, *Analyst*, 1944, **69**, 169.
[3] H. H. Willard and L. Gordon, *Analyt. Chem.*, 1948, **20**, 165.
[4] L. Gordon, C. H. Vanselow and H. H. Willard, *ibid.*, 1949, **21**, 1323.
[5] P. J. Elving and E. R. Caley, *Ind. Eng. Chem. Anal.*, 1937, **9**, 558.
[6] L. Gordon and E. R. Caley, *Analyt. Chem.*, 1948, **20**, 560.
[7] L. Gordon and A. F. Wroczynski, *ibid.*, 1952, **24**, 896.

6. PRECIPITATION OF PHOSPHATES

Zirconium is precipitated quantitatively from homogeneous solution by the slow hydrolysis of trimethyl phosphate which occurs on heating a solution in sulphuric acid.[1] A dense crystalline precipitate which is easily washed and filtered is obtained. The precipitate, which is slightly variable in composition, approximates to

$$ZrO[H(CH_3)PO_4]_2.2H_2O,$$

but contains varying amounts of zirconyl phosphate,

$$ZrO(H_2PO_4)_2.$$

On ignition, it is converted completely to zirconium pyrophosphate ZrP_2O_7, in which form it may be weighed.

By using controlled hydrolysis of triethyl phosphate in sulphuric acid solution, Willard and Freund[2] achieved fractional separation of zirconium and hafnium, probably as the ethyl phosphates,

$$MO[H(C_2H_5)PO_4]_2.2H_2O,$$

in spite of the close similarity between the two elements. Starting with 215 g. combined oxides containing 16·0 per cent by weight of hafnia, a series of five fractional precipitations yielded 7·16 g. of a fraction containing 91·1 per cent hafnia.

Preliminary experiments suggested that trimethyl phosphate, because of its more rapid breakdown, would speed up the procedure considerably. This would have a marked advantage, since each fractionation required 20 hours for complete precipitation.

In separating zirconium from elements other than hafnium, trimethyl phosphate can be used over a range of 2–60 mg. zirconium oxide. Above this content low results are obtained, probably arising from hydrolysis of the precipitate.

A precipitate which is not quite so good in character as that obtained by hydrolysis of the organic phosphate, but which is very much superior to that obtained by normal precipitation with orthophosphate, results from hydrolysis of metaphosphoric acid in homogeneous solution at room temperature. A procedure for the determination of zirconium based on this hydrolysis has been worked out.

The optimum concentration of acid is 3·6 N sulphuric acid (10 per cent by volume of the concentrated acid), but if antimony

3

or bismuth are present hydrochloric acid at about the same normality should be used. Up to 200 mg. of zirconium oxide may be separated quantitatively in the presence of aluminium, arsenate, borate, cadmium, cerium-III, chromium, cobalt, copper, magnesium, manganese, mercury-II, nickel, potassium, sodium, tartrate, vanadium-IV, yttrium and zinc. More than 0·025 g. iron-III, 0·010 g. thorium, or 0·020 g. titanium-IV oxide interferes. Antimony, bismuth, perchlorate and tin-IV interfere. Barium, lead and large amounts of calcium should be removed before carrying out the separation.

The procedure can be modified successfully to give good results in the presence of amounts of iron-III greater than 0·025 g. or in the presence of antimony or bismuth. By carrying out a double precipitation amounts of titanium-IV oxide up to 0·25 g. and of thorium up to 0·050 g. may be present.

Procedures

Separation of zirconium in the absence of interfering elements. If necessary remove barium, lead and large amounts of calcium by precipitation with sulphuric acid. Evaporate the solution with sulphuric acid to remove excess of other acids. Adjust the total volume of solution to 150–200 ml., and make 3·6 N in sulphuric acid. Cool the solution, and add 5 g. metaphosphoric acid dissolved in 25 ml. water. If titanium is present add 5 ml. 30 per cent hydrogen peroxide. Set aside for about 12 hours at room temperature. Bring the solution to just below boiling point, and maintain it at this temperature for 1 hour. Allow the solution to cool to room temperature; filter through a Whatman No. 40 filter paper, and wash with 300–500 ml. cold 5 per cent ammonium nitrate solution. Transfer filter paper and precipitate to a crucible, char the paper slowly, and ignite the zirconyl phosphate to constant weight at 900°–950° C. in a muffle. Weigh the pure white precipitate as ZrP_2O_7.

$$mg. \; ppt. \times 0·3440 \equiv mg. \; Zr.$$

Separation of zirconium in the presence of antimony. Prior to the precipitation add 10 g. tartaric acid to the sulphuric acid solution to prevent precipitation of the antimony. Otherwise proceed as in the normal precipitation.

Separation of zirconium in the presence of antimony and bismuth. Carry out the precipitation from a solution which, instead of containing sulphuric acid, is 3 N in hydrochloric acid. Use 200 ml. 5 per cent hydrochloric acid containing 4 g. ammonium phosphate to wash the precipitate, followed by 100 ml. 5 per cent ammonium nitrate solution.

Separation of zirconium in the presence of iron. If more than 25 mg. iron are present, remove the bulk of it by extraction with ether. Carry out the zirconium precipitation as described above, but after setting aside for 12 hours in the cold, maintain for 2 hours just under boiling temperature to allow any precipitated ferric metaphosphate to redissolve. Complete the determination as before.

Separation of zirconium in the presence of larger amounts of titanium or thorium. Carry out a precipitation in the normal manner, and wash the precipitate with 5 per cent ammonium nitrate solution. Return the filter paper and precipitate to the original beaker, and add 20 ml. concentrated nitric acid followed by 20 ml. concentrated sulphuric acid. Cover with a clock glass and heat until the filter paper is completely destroyed. Raise the clock glass and evaporate to dense white fumes. Cool, and dilute the solution to 200 ml. with water. Add 5 ml. 30 per cent hydrogen peroxide and 5 g. metaphosphoric acid dissolved in 25 ml. water. Complete the determination in the normal manner.

[1] H. H. Willard and R. B. Hahn, *Analyt. Chem.*, 1949, **21**, 293.
[2] H. H. Willard and H. Freund, *Ind. Eng. Chem. Anal.*, 1946, **18**, 195.

7. PRECIPITATION OF BENZOATES

After examining a number of reagents as possible alternatives to ammonia for the precipitation of the hydroxide analytical group, Kolthoff, Stenger and Moskowitz[1] concluded that ammonium benzoate was superior to any other precipitant, in that it allowed complete precipitation of aluminium, chromium and iron-III at lower values of pH, and yielded precipitates which were more readily filtered than those obtained with other reagents.

In dilute acetic acid solution the reagent gives a voluminous precipitate immediately with aluminium (white) and with iron-III (orange-tan), but 5 minutes' gentle boiling is required for complete precipitation. The chromium precipitate (grey-green) does not appear till the solution is heated, and 20 minutes' boiling is required for complete precipitation. The chromium and iron precipitates filter rapidly. The aluminium precipitate, while filtering more slowly, is still much superior to the hydroxide, filtering in about half the time required for the latter. The precipitates, which contain considerable amounts of benzoic acid, probably adsorbed, are rendered colloidal by washing with water, and so should be washed with benzoate solution. At higher pH values (pH 5·0–6·0) they are readily filtered.

After removal of these elements the normal qualitative scheme may be continued, as benzoate does not interfere. Phosphate is, however, only partially removed.

For separation of the group quantitatively, or for subsequent determination, the ions are first oxidised by hydrogen peroxide in alkaline medium. Aluminium and chromium (as chromium-VI) go into solution. The excess hydrogen peroxide is removed by boiling, and the iron-III precipitate may be filtered off and determined titrimetrically. After acidification, aluminium may be separated from the chromate by repeating the benzoate precipitation, and may then be determined in ammoniacal tartrate solution with 8-hydroxy-quinoline. Chromate may be determined titrimetrically in the filtrate.

Kolthoff and his co-workers state that, if present, ammonium, barium, cadmium, cerium-III, cobalt, iron-II, lithium, magnesium, manganese, mercury-II, nickel, potassium, sodium, strontium, vanadium-IV, vanadium-V and zinc are not precipitated under the conditions described. Partial precipitation of beryllium, copper, lead, tin-II, titanium-III and uranium occurs. Bismuth, cerium-IV, tin-IV, titanium-IV and zirconium are completely precipitated in the same conditions as used for the precipitation of the three elements under consideration.

When investigating the method for use in the separation of aluminium from some of the metals stated to remain in solution, Smales[2] concluded that the conditions laid down by Kolthoff and his co-workers were too imprecise for quantitative separation. This was particularly so with respect to pH control. Investigation showed that at pH 2·5–3·5 benzoic acid crystallises out during precipitation and gives rise to difficulties. At pH 4·0–5·0 the aluminium precipitate is granular and easily filtered. At pH 5·5–7·0 the precipitate is gelatinous and difficult to filter. As precipitation is quantitative from pH 3·5 upwards, the optimum pH range is 3·5–5·0, and preferably pH 4·0 should be chosen, as at this level adsorption of other ions is likely to be at its lowest. Bromophenol blue (pH 3·0–3·6) is recommended for indicating the correct pH.

Simple addition of ammonia to solutions containing ammonium chloride and ammonium benzoate does not permit precise enough control of the pH, so in addition Smales recommends that the solution should be buffered with ammonium acetate. Using these precautions, quantitative precipitation of aluminium in a granular readily filterable form is achieved. Separation of aluminium from up to at least 100 mg. chromium-VI or cobalt can be carried out by a single precipitation, but a double precipitation is required for separation from nickel or zinc.

Osborn and Jewsbury,[3] using precise pH control, have been able to obtain nearly quantitative separation of beryllium and aluminium. They confirm that in separate solution aluminium is completely precipitated at pH 3·5–4·0. Beryllium does not begin to precipitate until pH 6·6. Nevertheless, in a solution containing both elements maintained at pH 3·5–4·0, co-precipitation of beryllium to an extent of about 5 per cent occurs. By carrying out a reprecipitation, however, a useful separation can be obtained. Subsequent determination of the aluminium can be carried out over a wide range of concentrations with an error not greater than 2 per cent; and if the proportion of alumina to beryllia is not higher than 1 : 1, the beryllium can be determined with an error not exceeding 2 per cent. With a higher Al_2O_3 : BeO ratio, the beryllium determination becomes increasingly inaccurate.

Bayley[4] applied the benzoate method to the determination of aluminium in brasses, copper and iron being first reduced by hydroxylamine hydrochloride. The aluminium benzoate was subsequently redissolved and reprecipitated by 8-hydroxyquinoline for final determination.

Milner and Townend[5] found that Bayley's method, in spite of careful buffering, was not always successful for amounts of aluminium greater than 1 per cent, and that it could not be applied generally to aluminium bronzes or other alloys containing variable amounts of copper. The buffering capacity is insufficient to deal with the progressively increasing acidity of the solution which occurs as reduction of the copper by the hydroxylamine hydrochloride proceeds.

They have therefore modified Bayley's method to make it applicable to all forms of copper base alloy. By the new procedure 0·1–12·0 per cent of aluminium may be determined in samples of coppers, brasses and aluminium bronzes containing up to 0·5 g. copper.

The separation of aluminium from iron-III by the benzoate method is possible if the iron is first reduced, since iron-II does not precipitate. For reduction purposes thioglycollic acid, which then forms a soluble complex with the iron-II, has been recommended.[6] Up to 1 g. iron may be present, and very large amounts of salts such as ammonium chloride, ammonium sulphate, ammonium perchlorate and sodium chloride do not interfere. Bivalent metals, tungstate and molybdate do not interfere in the separation. In the presence of phosphate the aluminium is precipitated as aluminium phosphate, free from iron. Chromium, titanium and vanadium interfere.

Following on the use of thioglycollic acid as a reducing agent, Jewsbury and Osborn[7] have investigated other elements which give insoluble benzoates, and the effects of various complexing agents such as tartaric acid, citric acid and salicylic acid in addition to thioglycollic acid.

With no complexing agent present, and using hydrochloric acid solution, aluminium, beryllium, bismuth, cerium-IV (sulphuric acid solution), chromium-III, iron-III, thorium, tin-II, tin-IV, titanium-III, titanium-IV, uranium and zirconium were found to give insoluble benzoates. Ammonium, barium, cadmium, calcium, cobalt, lithium, magnesium, manganese, mercury-II, nickel, potassium, sodium, strontium, vanadium-IV, vanadium-V and zinc gave no precipitate. Copper gave no precipitate in the presence of acetate, antimony gave no precipitate if nitric acid was used instead of hydrochloric acid, and lead gave no precipitate up to pH 7·0 and only a turbidity beyond that point.

In the presence of thioglycollic acid, aluminium, beryllium, hafnium, thorium, titanium-III, titanium-IV and zirconium gave insoluble benzoates, but not bismuth, cerium-IV, chromium-III, iron-III, tin-II, tin-IV or uranium. With salicylic acid, which also showed selective complexing action, titanium-III and titanium-IV precipitated completely at pH 7·0. Bismuth precipitated over the pH range 2·5–7·0. Tin-II precipitated over the pH range 3·0–7·0. Cerium-IV, hafnium, thorium and zirconium precipitated as salicylates and not as benzoates. Aluminium, beryllium, chromium and iron-III did not precipitate.

Based on these observations, Osborn and Jewsbury have proposed methods for the separation of bismuth, thorium, tin-IV, titanium and zirconium. It was not, however, found possible to devise a satisfactory method for cerium.

In the separation of tin, the elements aluminium, bismuth, cerium-IV, chromium-III, iron-III, thorium and zirconium are also precipitated, and must be absent. Antimony, copper and lead co-precipitate sufficiently in a single precipitation to produce an appreciable error. The separations of thorium, titanium and zirconium are carried out using thioglycollic acid as complexing agent, the only other elements precipitated in these conditions being aluminium and hafnium. A good separation of titanium from calcium, chromium-III, iron, magnesium and tin is obtained by a double precipitation. Oxalate interferes with the titanium precipitation. Separation of titanium from aluminium may be achieved by the addition of salicylic acid before precipitation of the benzoate, and it is suggested that this procedure might also be applied to the removal of hafnium, thorium and zirconium before precipitating titanium (p. 27). Copper is co-precipitated with zirconium in the presence of thioglycollic acid. If, therefore, copper is present, the complexing agent is omitted. Separation of copper from titanium is not possible.

Normally thorium does not precipitate in the presence of sulphuric acid, and this precipitation should be carried out in nitric or hydrochloric acid solution. A good separation of thorium from cerium, iron and yttrium is found, and the benzoate method is therefore suggested as a possible means of separating thorium from the lanthanons and from scandium.

In the precipitation of bismuth, hydrochloric acid and acetate interfere. Nitric acid solution must therefore be used. The precipitation is quantitative, but a good separation from lead is not obtained, and the precipitate does not ignite satisfactorily to the oxide. Precipitation of bismuth is not, therefore, recommended as a quantitative procedure.

Solutions required

Reagent solution. Dissolve 100 g. ammonium benzoate, preferably purified by recrystallisation from dilute ammonia solution, in 1 litre water with warming. Add 1 mg. thymol to prevent mould growth. Cool and filter if necessary. This solution is quite stable in glass containers.

Wash solution I. Dilute 100 ml. reagent solution to 1 litre with water.

Wash solution II. Dilute 100 ml. reagent solution and 20 ml. glacial acetic acid to 1 litre with water. Benzoic acid separates out at room temperature. Warm, and use the warm solution for washing.

Wash solution III. Dissolve 10 g. ammonium benzoate in 1 litre water. Adjust the pH (with hydrochloric acid and ammonia) to the full yellow of thymol blue (pH 3·0).

Wash solution IV. Dissolve 10 g. ammonium benzoate in 1 litre water. Adjust the pH to the orange of cresol red (pH 7·5).

Wash solution V. Dissolve 10 g. ammonium benzoate in 1 litre water. Adjust the pH (with nitric acid and ammonia) to the first yellow of cresol red (pH 1·8).

Buffer solution. Dissolve 75 g. powdered anhydrous sodium acetate in 640 ml. N hydrochloric acid.

Ammoniacal tartrate solution. Dissolve 25 g. tartaric acid in water. Add 120 ml. concentrated ammonia (sp. gr. 0·88) and 5 g. potassium cyanide, and dilute to 1 litre with water.

Ammonium acetate solution. A 10 per cent aqueous solution prepared from AnalaR acetic acid and AnalaR ammonium hydroxide to avoid aluminium impurity.

8-Hydroxyquinoline solution. Dissolve 5 g. 8-hydroxyquinoline in 15 ml. glacial acetic acid and dilute to 250 ml. with water.

Procedures

Precipitation of aluminium, chromium and iron-III. Oxidise the hydrochloric acid solution to convert iron-II to iron-III, and dilute to about 100 ml. with water. Treat with dilute ammonia solution until the precipitate which is formed only redissolves very slowly on stirring. Add 1 ml. glacial acetic acid, and enough ammonium chloride to produce a total content of at least 1 g. of this salt. Add 20 ml. reagent solution for every 65 mg. aluminium or 125 mg. iron or chromium present. Heat the suspension, while stirring, till boiling commences, and keep it gently boiling for 5 minutes in the absence of chromium, or for 20 minutes if appreciable amounts of that element are present. Filter through a Whatman No. 1 filter paper. Wash the precipitate ten times with hot wash solution II. For exact work evaporate the filtrate and washings to 50 ml., and filter to remove unprecipitated chromium or iron.

Precipitation of aluminium. Dilute the solution, which should be slightly acid with hydrochloric acid, to 250–300 ml. with water. Add 1 g. ammonium chloride, 20 ml. 10 per cent ammonium acetate solution, 20 ml. reagent solution and 2 ml. bromophenol blue indicator solution. Heat to 80° C. to dissolve benzoic acid, and add sufficient hydrochloric acid to dissolve any precipitated aluminium benzoate. Add dilute ammonia solution slowly from a burette, shaking the solution continually, until the indicator just begins to change colour and precipitation just begins. Boil for 1–2 minutes. Continue addition of the dilute ammonia from the burette to counteract the increasing acidity and to turn the indicator red-blue (pH 3·5–4·0). Boil gently for 2–3 minutes, and place on a water-bath for half an hour to allow the precipitate to settle. Filter through a Whatman No. 40 filter paper, and wash the precipitate ten times with hot wash solution I. Ignite the precipitate to constant weight, first heating gently in a muffle, and then over a blast lamp. Weigh as Al_2O_3.

$$mg. \text{ ppt.} \times 0·5291 \equiv mg. \text{ Al.}$$

Separation of aluminium from copper base alloys. Dissolve the sample of alloy, containing 0·005–0·01 g. aluminium, in 5 ml. concentrated nitric acid (sp. gr. 1·20) in a 400-ml. conical beaker. Add approximately 10 ml. water and boil to drive off nitrous fumes. If tin is absent dilute to approximately 75 ml. with water. In the presence of tin, evaporate to a paste, add 15 ml. 0·5 per cent nitric acid and filter through a small, tight paper-pulp pad. Wash this well with approximately 50 ml. hot 0·5 per cent nitric acid, and collect filtrate and washings in a 400-ml. conical beaker.

To the solution free from tin add carefully 1 : 1 ammonia solution to the point where the first permanent precipitate is produced, followed by 1 : 4 hydrochloric acid, added dropwise to clear the solution. Add 70 ml. buffer solution and 15 ml. 5 per cent hydroxyl-amine hydrochloride solution, and heat to boiling. Boil for 1 minute, remove from the hot plate, and add 20 ml. reagent solution in one rapid addition. Place the vessel by the side of the hot plate for about 15 minutes to allow the precipitate to settle and the supernatant liquid to become perfectly clear. Filter through a moderately large paper-pulp pad of medium compactness, and wash well with hot wash solution I.

Redissolve the aluminium benzoate in 50 ml. hot ammoniacal tartrate solution. Dilute to 150 ml. with water, digest at 80°–90° C. for a few minutes, and precipitate the aluminium by adding 20 ml. 8-hydroxyquinoline solution with constant shaking. Filter off the precipitate, and determine titrimetrically by bromate-bromide titration.

1 ml. N potassium bromate≡0·2248 mg. Al.

Separation of aluminium from iron. Dilute the solution, from which silica has been removed in the normal way, to about 200–300 ml. in a 500-ml. beaker. Nearly neutralise with ammonia. Add sufficient ammonium chloride to give a total content of 5 g. ammonium salts. Add 1 ml. 90 per cent thioglycollic acid, 20 ml. ammonium acetate solution, 20 ml. reagent solution, and 2 ml. bromophenol blue indicator solution. Heat to about 80° C., when any precipitated benzoic acid will redissolve. Redissolve any precipitated aluminium benzoate by dropwise addition of dilute hydrochloric acid. From a burette add 1 : 4 ammonia solution, stirring thoroughly, till the indicator begins to change colour and precipitation just begins. Boil for 2–3 minutes, and if necessary add more ammonia and boil for a further minute. Continue adding ammonia to the boiling solution till the colour changes to red-blue (pH 3·5–4·0). Boil gently for 2–3 minutes, allow to settle on a boiling water-bath for half an hour, and filter through a Whatman No. 40 filter paper. Wash ten times with hot wash solution II.

If the amount of iron is very large, wash only two to three times, redissolve the precipitate in hot dilute hydrochloric acid, add a few grams ammonium chloride, and reprecipitate. Wash this precipitate ten times.

Transfer the paper and precipitate to a silica or platinum crucible, char in front of the muffle, and ignite finally at about 1000° C. Weigh as Al_2O_3.

mg. ppt. ×0·5291≡mg. Al.

Precipitation of tin-IV. Dissolve the sample in 20 ml. concentrated hydrochloric acid and oxidise with bromine, as tin-II is not completely precipitated. Remove excess bromine by boiling, and dilute to 150 ml. Add 20 ml. reagent solution and 20 ml. ammonium acetate solution, followed by a few ml. thymol blue indicator solution. Heat almost to boiling and add strong ammonia solution, drop by drop, with stirring, until the indicator changes to full yellow (pH 3·0). Boil for 2 minutes, place on a steam bath for 15 minutes, and filter through a Whatman No. 41 filter paper. If a little of the precipitate passes through the paper at the beginning of the filtration, refilter the first few ml. (A precipitate which forms after filtration will be benzoic acid crystallising out on cooling.) Wash the precipitate with 200 ml. wash solution III, and then with 200 ml. hot 2 per cent ammonium nitrate solution to remove benzoic acid.

Dry the filter and precipitate and ignite carefully in a silica crucible, heating with nitric acid before the final ignition. Weigh as SnO_2.

$$\text{mg. ppt.} \times 0.7877 \equiv \text{mg. Sn.}$$

Precipitation of titanium. Carry out the precipitation exactly as described for tin, but adjust the pH to 3·5 by using bromophenol blue as indicator. Add an excess of thioglycollic acid before precipitating with ammonia. Wash the precipitate with hot wash solution I. Ignite in a platinum crucible. Weigh as TiO_2.

$$\text{mg. ppt.} \times 0.5995 \equiv \text{mg. Ti.}$$

Precipitation of titanium in the presence of aluminium. Add 10 g. salicylic acid for every 0·1 g. aluminium oxide, adjust the pH to 7·5 before precipitation, and wash with hot wash solution IV.

Precipitation of zirconium. Dilute the solution to 150 ml. and add 30 ml. concentrated hydrochloric acid, 20 ml. reagent solution and 20 ml. ammonium acetate solution. In the presence of interfering elements (cerium, iron, tin) add excess of thioglycollic acid. Heat to boiling, and add 20 drops 0·02 per cent cresol red solution and 7 drops 0·1 per cent methyl violet 6B solution. Add strong ammonia solution with constant stirring until the indicator (which is partly absorbed on the precipitate) turns to a full blue (pH *c.* 1·5). Boil for 2 minutes, and place on a steam bath for 15 minutes. Filter through a Whatman No. 41 filter paper, refiltering the first few ml. of filtrate.

Wash with boiling wash solution IV. Ignite the precipitate to the oxide in a platinum crucible. Weigh as ZrO_2.

$$mg. \text{ ppt.} \times 0.7403 \equiv mg. \text{ Zr.}$$

Precipitation of zirconium in the presence of copper. Carry out the precipitation as described for zirconium, but without the addition of thioglycollic acid. Redissolve the precipitate and reprecipitate, this time adding thioglycollic acid.

Precipitation of thorium. Bring the volume of solution to 150 ml., and add 10 ml. concentrated nitric acid. Heat almost to boiling, and add 20 ml. reagent solution, 20 ml. ammonium acetate solution and 3 ml. 90 per cent thioglycollic acid solution. Add a few drops of cresol red and then add ammonia solution, drop by drop, until the solution just turns yellow (pH 1·8). Boil for 2 minutes, place on the steam-bath for 15 minutes, and filter through a Whatman No. 41 filter paper. Wash with wash solution V, dry and ignite. Weigh as ThO_2.

$$mg. \text{ ppt.} \times 0.8789 \equiv mg. \text{ Th.}$$

Precipitation of bismuth. Acidify the solution containing bismuth nitrate with nitric acid, and adjust the pH to the full yellow of thymol blue (pH 3·0). Precipitate with reagent solution, adding no acetate, and filter through a Whatman No. 41 filter paper.

[1] I. M. Kolthoff, V. A. Stenger and B. Moskowitz, *J. Amer. Chem. Soc.*, 1934, **56**, 812. Cf. also *Qualitative Analysis and Analytical Chemical Separations*, P. W. West, M. M. Vick and A. L. Le Rosen, New York, 1953, 26 ff.
[2] A. A. Smales, *Analyst*, 1947, **72**, 14.
[3] G. H. Osborn and A. Jewsbury, *Anal. Chim. Acta*, 1949, **3**, 108.
[4] W. J. Bayley, *Chem. and Ind.*, 1950, 34.
[5] G. W. C. Milner and J. Townend, *Analyst*, 1951, **76**, 424.
[6] H. N. Wilson, *Anal. Chim. Acta*, 1947, **1**, 330.
[7] A. Jewsbury and G. H. Osborn, *ibid.*, 1949, **3**, 642.

8. PRECIPITATION OF TETRACHLOROPHTHALATES

As already stated, the basic formate method for the separation of thorium from the lanthanons is not wholly satisfactory (p. 10). Since thorium is precipitated by dibasic organic acids other than oxalic acid and succinic acid, Gordon, Vanselow and Willard[1] have studied the action of a number of these, used as precipitants in homogeneous solution, for the purpose of separating thorium from the lanthanons after precipitation of these elements by oxalate hydrolysis (p. 13). They have reported favourably on tetrachlorophthalic acid,

With this reagent no precipitate is formed at room temperature, even after several hours; at or near the boiling point a gelatinous precipitate is produced. However, gradual warming to 70°–80° C., with constant stirring, produces a dense crystalline precipitate which is easily filterable.

The authors conclude that in this precipitation the mechanism is different in type from that occurring in urea hydrolysis or that occurring in ester hydrolysis. They suggest that it may be due to increased dissociation of a complex of thorium with hydroxyl as the temperature rises, followed by combination with the reagent to give the desired precipitate.

The precipitate is slightly variable in composition, and is apparently largely a basic salt which corresponds to one thorium ion, two tetrachlorophthalate ions and three molecules of water. It must therefore be ignited to the dioxide for subsequent determination.

The optimum pH for precipitation of thorium is 1·0. At pH 2·0, cerium-III is also precipitated by the reagent. Below this value lanthanum, cerium-III, praseodymium, neodymium and yttrium are not precipitated, so that thorium may be separated from these elements. When the method is applied to the separation of thorium from monazite sand extracts, a double precipitation is recommended because of the large amount of lanthanons present. The thorium and lanthanons are first extracted by the procedure already described (p. 14).

Procedure

Separation of thorium from lanthanons. To the solution of perchlorates in a 600-ml. beaker add 200 ml. water and 1 g. sodium iodide. Adjust the pH to 1·5–1·6. Add 200 ml. tetrachlorophthalic acid solution containing 3 g. per litre, which is 2·5 times the theoretical amount required for precipitation of 0·1 g. thorium oxide. Adjust the pH to 1·0–1·1. Maintain at 70°–80° C. on a hot plate, and stir mechanically, continuing this for 1½ hours after the initial appearance of the precipitate (20 minutes for 50–100 mg. of thorium oxide, 1–2 hours for amounts much less than this).

Filter the hot solution, and wash with a cool 0·1 per cent solution of the reagent adjusted to pH 1·5 with hydrochloric acid. Convert the precipitate to insoluble thorium hydroxide by treating the tetrachlorophthalate precipitate with hot 2 per cent sodium hydroxide, washing at least ten times with this reagent to remove all the tetra-chlorophthalic acid as the sodium salt. Dissolve the hydroxide in hot 2 N hydrochloric acid, receiving the solution in the original beaker.

Reprecipitate the thorium with tetrachlorophthalic acid as before. Transfer the precipitate to a porcelain filter crucible and wash with dilute reagent solution (pH 1·5) as before. Dry at 110° C., place the crucible in a muffle, and bring the temperature gradually to 350° C. Ignite at this temperature for 45 minutes, and then at 850° C. for 1 hour. Weigh as ThO_2.

$$mg. ppt. \times 0·8789 \equiv mg. Th.$$

[1] L. Gordon, C. H. Vanselow and H. H. Willard, *Analyt. Chem.*, 1949, **21**, 1323.

9. PRECIPITATION OF BARIUM

Aqueous solutions of sulphamic acid, on heating, undergo hydrolysis with the production of sulphate ion:

$$NH_2 . SO_3H + H_2O = NH_4HSO_4.$$

Willard[1] has used this reaction as a means of separating calcium from barium. It has also been recommended for the precipitation of barium in homogeneous solution.[2] The resulting precipitate of barium sulphate is coarsely crystalline, filtering readily. It contains few co-precipitated impurities, and is readily washed free from adsorbed ions. Considerable amounts of calcium, iron, magnesium, sodium, nitrate and phosphate cause little or no interference, but strontium in anything other than trace amounts interferes.

Procedure

Precipitation of barium. Dilute the barium solution to approximately 100 ml. Add 1 g. solid sulphamic acid, and place on a steam bath. Heat for about 30 minutes after the first turbidity appears. Filter through a weighed porcelain filter crucible, wash with hot water, and ignite to constant weight at 900° C. or above. Weigh as $BaSO_4$.

$$\text{mg. ppt.} \times 0.5885 \equiv \text{mg. Ba.}$$

[1] H. H. Willard, *Analyt. Chem.*, 1950, **22**, 1372.
[2] W. F. Wagner and J. A. Wuellner, *ibid.*, 1952, **24**, 1031.

CHAPTER II

SEPARATION BY EXTRACTION

EXTRACTION processes have been subjected to a considerable renewal of interest and to extension of their applications. This arises partly from the realisation that separation by using differences in partition coefficients is readily susceptible to the application of continuous or countercurrent methods; and partly because of the wide range of covalent organo-metallic compounds now known, many of which are readily soluble in organic liquids and comparatively insoluble in water.

In consequence of this, extraction processes which aim at quantitative separation rather than processes which aim at the removal of the bulk of an interfering or major constituent, will undoubtedly become known as a major analytical technique. While the theoretical principles underlying such separations have not, as yet, been fully developed, they have received some attention.[1] Bearing in mind the scope of the present work, however, the separations by extraction which are dealt with here will be approached largely from the classical standpoint.

A. EXTRACTION OF INORGANIC SALTS

The extraction of inorganic salts from aqueous or acid solution by means of an immiscible organic liquid is a classical procedure, the best known and most widely practised instance being the use of diethyl ether for the extraction of a wide range of salts,[2] and in particular, for the extraction of large amounts of iron-III as ferric chloride before determining other cations.

More recent investigations have extended the list of organic solvents which may be used, so that a number of ethers, esters and alcohols are now known to be suitable for certain extractions.[3]

[1] H. M. Irving, *Quart. Reviews*, 1951, **5**, 200.
[2] G. H. Morrison, *Analyt. Chem.*, 1950, **22**, 1388.
[3] J. W. Rothe, *Stahl. u. Eisen*, 1892, **12**, 1052: 1893, **13**, 333: *Mitt. könig. tech. ver.*, 1892, **10**, 132: *Chem. News*, 1892, **66**, 182.

1. Diethyl Ether

$$(C_2H_5)_2O. \quad B.p. \ 34 \cdot 5°C. \quad D. \ 0 \cdot 713.$$

Extraction of uranium. Uranyl nitrate in a solution more acid than pH 4·0, is extracted by diethyl ether.[1] The extraction is favoured by the presence of excess nitrate ion, preferably present as ferric nitrate. It is also favoured by increase in the acid concentration, but this may lead to the extraction of other salts.

Scott[2] has shown that optimum conditions exist when the solution is 3 N in nitric acid and molar in ferric nitrate, 99 per cent of uranium being removed by three successive extractions with a volume of ether equal to the volume of solution. Thorium and nickel interfere with the extraction.

Helgar and Rynninger[3] have devised a procedure applicable to minerals. The material is powdered, heated at 600°C. to remove organic matter and sulphur, and fused with a mixture of sodium carbonate and sodium peroxide. The resulting melt is dissolved up and passed through an anion-exchange column to remove sulphate and phosphate. The effluent from the column is twice evaporated to fuming with nitric acid. The acidity is adjusted, the solution is saturated with ammonium nitrate, and it is then extracted with ether. The ether is evaporated off, and the uranium is finally precipitated with 8-hydroxyquinoline, ignited to U_3O_8, and redissolved in dilute nitric acid.

Saturation with ammonium nitrate before extraction has been recommended by several other authors.[4, 5]

[1] K. Lindth, R. Rynninger and E. Skoraeus, *Svensk Kem. Tidskr.*, 1949, **61**, 170.
[2] T. R. Scott, *Analyst*, 1949, **74**, 486.
[3] B. Helgar and R. Rynninger, *Svensk Kem. Tidskr.*, 1949, **61**, 189.
[4] F. Hecht and A. Grünwald, *Mikrochem. Mikrochim. Acta*, 1943, **30**, 279.
[5] C. J. Rodden, *Analyt. Chem.*, 1949, **21**, 327.

2. Di-*iso*-Propyl Ether

$$[(CH_3)_2CH]_2O. \quad B.p. \ 69°C. \quad D. \ 0 \cdot 725.$$

Extraction of iron. Di-*iso*-propyl ether, first recommended for the extraction of iron by Mellor,[1] has been fully investigated, and is claimed to be a more satisfactory extractant for iron-III than diethyl ether.[2] In its favour are its higher efficiency under comparable conditions, its much lower solubility in the aqueous phase, and its lower inflammability.

At room temperature the best extraction is obtained from a solution which is 7·5–8·0 N in hydrochloric acid, although good extraction is obtained over a fairly wide range of acidity—wider than for diethyl ether. The percentage extraction varies inversely with the amount of iron present. With a solution containing 1 mg. iron, 96 per cent extraction results (the corresponding figure for diethyl ether being 80 per cent) and extraction is virtually complete after three successive extractions.

Solutions to be extracted should first be taken to dryness with an oxidising agent to oxidise ferrous salts. They should then be converted completely to chlorides, and dissolved in 8 N hydrochloric acid.

Separation of iron-III from aluminium, chromium-III, copper, manganese, nickel, titanium-IV, vanadium-IV and zinc is complete. Molybdenum, phosphate and large amounts of vanadium-V pass into the organic layer.

Although the authors claim that normal impurities in the ether (*iso*-propyl alcohol, peroxides) do not affect iron-III, Ashley and Murray[3] state that reduction to the inextractable iron-II form may be caused by these or may take place photochemically. They recommend extraction in the absence of strong sunlight.

[1] J. W. Mellor, *Trans. Eng. Ceram. Soc.*, 1910, **8**, 125, 132.
[2] R. W. Dodson, G. J. Forney and E. H. Swift, *J. Amer. Chem. Soc.*, 1936, **58**, 2573.
[3] S. E. Q. Ashley and W. M. Murray, *Ind. Eng. Chem. Anal.*, 1938, **10**, 367.

Extraction of antimony. Antimony-V is preferentially extracted from antimony-III in hydrochloric acid (above 10 per cent) by diethyl ether.[1, 2] The separation is not efficient, some antimony-III also being extracted, and some antimony-V being left in the aqueous phase. Thus, in 6 N hydrochloric acid, 81 per cent of the quinquivalent and 6 per cent of the tervalent forms are extracted.

Using di-*iso*-propyl ether and a hydrochloric acid range of 6–9 N, it has been found that extraction of antimony-V is rapid and effectively complete.[3] A small amount of antimony-III is also transferred to the ether layer, but a single washing of this with 7–8 N hydrochloric acid removes all but 0·04 per cent of the tervalent form without re-extracting any of the antimony-V.

[1] F. Mylius and C. Hüttner, *Ber.*, 1911, **44**, 1315.
[2] E. H. Swift, *J. Amer. Chem. Soc.*, 1924, **46**, 2375.
[3] F. C. Edwards and A. F. Voigt, *Analyt. Chem.*, 1949, **21**, 1204.

3. β: β'-Dichlorodiethyl Ether,

$$(Cl.CH_2.CH_2)_2O. \quad B.p.\ 178°C. \quad D.\ 1·22.$$

Extraction of iron. The mutual solubilities of this solvent and hydrochloric acid are very small. The liquid is non-inflammable. It has therefore been recommended[1] as an extractant for iron. Since the specific gravity of concentrated hydrochloric acid is about 1·20, several successive extractions can be carried out in an ordinary separating funnel, and there is no necessity for a special extraction apparatus.

The solution to be extracted should be at least 7 N in hydrochloric acid, but the acidity may range from this up to concentrated acid without any loss of extracting power. Very little heating takes place, whereas in extractions with diethyl ether the amount of heat evolved is inconveniently large.

Extraction is slower than with diethyl ether or with di-*iso*-propyl ether, and has been stated to be somewhat less efficient. However, the reagent is claimed to be particularly useful for the extraction of high concentrations of iron.

[1] J. Axelrod and E. H. Swift, *J. Amer. Chem. Soc.*, 1940, **62**, 33.

4. Ethylene Glycol Monobutyl Ether (Butyl Cellosolve)

$$CH_2OH.CH_2.O.C_4H_9. \quad B.p.\ 171°C. \quad D.\ 0·903.$$

Extraction of calcium. In the separation of the alkaline earth cations, use has been made of the solubilities of the chlorides or the nitrates in various organic solvents, after a preliminary separation of the group from other elements as carbonates. Fairly satisfactory separations have been obtained with diethyl ether-ethanol mixtures, or with *iso*-propyl alcohol, acetone, amyl alcohol or butyl alcohol. All these solvents must be used in strictly anhydrous conditions, and the ether-ethanol mixture has the additional disadvantage of being highly inflammable. In no case is the separation complete.

Butyl cellosolve has been recommended for the separation of calcium as nitrate from the other two alkaline earth elements.[1] It is immiscible with water and does not form a binary complex with it. It is therefore possible to dehydrate the nitrates and the solvent simultaneously, simply by boiling the aqueous phase together with the solvent until all the water has been driven off. Anhydrous calcium nitrate dissolves in the solvent to the extent of 0·243 g. per 100 ml. Barium and strontium nitrates are practically insoluble, their solubilities being of the same order as those of barium sulphate or strontium carbonate in water.

[1] H. H. Barber, *Ind. Eng. Chem. Anal.*, 1941, **13**, 572.

5. Ethyl Acetate

$$CH_3.COO.C_2H_5. \quad \text{B.p. } 77°C. \quad D. \ 0.900.$$

Extraction of gold. Gold may be extracted by esters from solutions which are about N in hydrochloric acid.[1] As the number of carbon atoms in the ester increases, the extractive power decreases, and ethyl acetate proves to be the most efficient extractant. The amount of gold extracted increases with acid concentration up to 10 per cent, and then remains roughly constant. At 30 per cent acid the two layers become completely miscible.

Ethyl acetate has been utilised for the separation of gold from palladium.[2] The solution, containing gold-III and palladium-II chlorides (at a concentration of about 10 per cent hydrochloric acid) is shaken with ethyl acetate in a separating funnel. The aqueous phase is drawn off and the organic phase is washed with a few ml. N hydrochloric acid.

Ethyl acetate is stated to be a more efficient extractant for gold than diethyl ether, which has also been recommended,[3] fewer extractions by the ester being required for almost complete removal of the gold.

Like palladium, aluminium, antimony-III, arsenic (AsO_4'''), bismuth, barium, cadmium, calcium, chromium-III, cobalt, copper, iron-III, lead, magnesium, manganese-II, mercury-II, nickel, potassium, sodium, strontium, tin-IV and zinc are not extracted appreciably under these conditions. In the presence of copper, lead, potassium or sodium, however, double salts are formed which lower the extractability of the gold.

[1] V. Lenher and C. H. Kao, *J. Phys. Chem.*, 1926, **30**, 126.
[2] J. H. Yoe and L. G. Overholser, *J. Amer. Chem. Soc.*, 1939, **61**, 2058.
[3] F. Mylius, *Z. anorg. Chem.*, 1911, **70**, 203.

6. Amyl Acetate

$$CH_3 . COO(CH_2)_2CH : (CH_3)_2.$$ B.p. *c.* 140°C. Sp. gr. 0·88.

Extraction of iron. In a single extraction of iron from concentrated hydrochloric acid solutions by amyl acetate, 99·6 per cent of the iron passes into the organic layer.[1] Addition of concentrated sulphuric acid or phosphoric acid to the hydrochloric acid solution renders extraction effectively complete. It is claimed that extraction is more efficient and acid concentration less critical than when ethers are used. Large amounts of molybdenum (82 per cent), tin-IV (68 per cent) and vanadium-V (14 per cent), and trace amounts of aluminium, cobalt and titanium-IV are also extracted.

Since the density of the organic liquid is less than that of the aqueous phase, a suitably designed extraction apparatus is desirable.

[1] J. E. Wells and D. P. Hunter, *Analyst*, 1948, **73**, 671.

7. 2-Ethyl Hexanol

$$CH_3(CH_2)_2 . CH . (C_2H_5)CH_2OH.$$ B.p. 184°C. D. 0·834.

Extraction of lithium. For the extraction of lithium chloride from sodium and potassium chlorides, acetone, amyl alcohol, dioxan and pyridine have all been proposed. Of these, amyl alcohol, recommended by Gooch,[1] seems to have been the most widely used. It possesses the advantage that the aqueous layer can be evaporated by boiling, thus carrying out the necessary dehydration and taking up the lithium chloride while allowing the other chlorides to precipitate out slowly. However, sodium and potassium chlorides are not completely insoluble in amyl alcohol, and in any subsequent determination of lithium a correction must be applied.

The solubilities of the chlorides decrease progressively with increase in molecular weight of the alcohol, and Caley and Axilrod[2] have found that in *n*-hexanol or in 2-ethyl hexanol the solubilities of sodium and potassium chlorides are reduced to the point where they are negligible, while lithium chloride is still sufficiently soluble for extraction to be practicable.

In 2-ethyl hexanol, which is slightly more selective, potassium chloride is effectively insoluble (0·01 mg. per 100 ml.) and sodium chloride is only slightly soluble (0·1 mg. per 100 ml.). Lithium chloride is soluble to the extent of 3 g. per 100 ml.

If a considerable amount (*c.* 100 mg.) of lithium is present, and if dehydration is carried out at the boiling point of the solvent, low extraction results. This is due to conversion of appreciable amounts of the lithium chloride into the insoluble lithium hydroxide. If the operation is carried out at 135°C. complete extraction is achieved.

[1] F. A. Gooch, *Proc. Amer. Acad. Arts Sci.*, 1886, **22**, 177: *Chem. News*, 1887, **55**, 18, 29, 40, 56, 78.
[2] E. R. Caley and H. D. Axilrod, *Ind. Eng. Chem. Anal.*, 1942, **14**, 242.

8. Methyl *iso*-Propyl Ketone

$$CH_3.CO.CH(CH_3)_2.\quad B.p.\ 95°C.\quad D.\ 0·803.$$

Extraction of lead. Bismuth, cadmium, copper, gold, iron, lead, mercury, palladium, rhodium and ruthenium are partially or completely extracted from aqueous solutions containing iodide by a wide range of organic solvents.[1] West and Carlton have investigated the conditions for the extraction of lead. Methyl *iso*-propyl ketone was found to be particularly successful for this purpose, a single treatment under suitable conditions of a solution containing 119 μg. lead giving 97 per cent extraction.

Lead iodide is not completely taken up by the organic solvent from neutral solution, but a concentration of 5 per cent of concentrated hydrochloric acid (by volume) in the aqueous phase provides for almost complete extraction. A high concentration of iodide is essential.

Copper, gold, lead, mercury, palladium and zinc can be extracted from a solution containing thiocyanate, while antimony and tin can be extracted from a solution containing hydrochloric acid. Preliminary treatment with ammonium thiocyanate and hydrochloric acid, and an extraction prior to the addition of the iodide, eliminate interference from these elements. Aluminium, barium, calcium, chromium, cobalt, iridium, magnesium, nickel, osmium and strontium are not extracted under the conditions of the experiment. Arsenic is partially extracted. When the treatment was applied to solutions containing lead in conjunction with binary combinations of the elements antimony, bismuth, cadmium, copper, gold, mercury, palladium, platinum, rhodium, ruthenium and zinc, spectrographic examination showed that the only other elements to be extracted were cadmium and ruthenium.

Procedure

To the solution containing about 100 μg. of lead add an equal volume of saturated ammonium thiocyanate solution and sufficient hydrochloric acid to make the solution 5 per cent in this acid. Transfer the whole to a separating funnel, and extract with an equal volume of methyl iso-propyl ketone. Transfer the aqueous phase to another separating funnel, add 3·75 ml. saturated aqueous potassium iodide solution, and sufficient concentrated hydrochloric acid to bring the total volume of this acid in the solution to 1·25 ml. Dilute to 25 ml. with water, and shake with an equal volume of the solvent. Repeat the extraction with a further volume of the ketone and combine the two extracts. Liberate the lead from the organic phase and extract back.

[1] P. W. West and J. K. Carlton, *Anal. Chim. Acta*, 1952, 6, 406.

B. EXTRACTION OF ORGANOMETALLIC COMPLEXES

Many organometallic complexes are soluble in organic solvents, and cations may therefore be extracted from aqueous solution by use of the appropriate reagent and a suitable solvent. For most reagents so far investigated, the range of extractable cations is extensive, and the conditions are markedly dependent on such functions as pH.[1, 2]

Few of the extractive processes can, as yet, therefore, be said to have reached the same stage of general applicability to specific problems as have the inorganic extractions already quoted, although they have found use in limited instances. In many cases extraction is, of course, a preliminary to colorimetric or absorptiometric analysis. While the applications to inorganic analysis will undoubtedly develop widely in the near future, few of the methods are sufficiently specific to warrant inclusion here.

At least four systems have been the subject of considerable investigation, and the extractable cations for these are as follows:

1. 8-hydroxyquinolates[3, 4]

Aluminium, bismuth, cadmium, cobalt, copper, gallium, indium, iron, lead, manganese, molybdenum, nickel, palladium, scandium, thallium, thorium, tin, titanium, uranium, vanadium, zinc, zirconium.

2. Cupferrates[5]

Aluminium, antimony, bismuth, cerium, cobalt, copper, indium, iron, manganese, mercury, molybdenum, nickel, niobium, thorium, tin, titanium, tungsten, uranium, vanadium, zinc, zirconium.

3. Dithizonates[6]

Bismuth, cadmium, cobalt, copper, gold, indium, iron, lead, manganese, mercury, nickel, palladium, platinum, silver, thallium, tin, zinc.

4. Diethyldithiocarbamates[7]

Bismuth, cobalt, copper, chromium, iron, molybdenum, nickel, tin, uranium.

[1] H. M. Irving and R. J. P. Williams, *J.C.S.*, 1949, 1841.
[2] H. M. Irving, *Quart. Reviews*, 1951, 5, 200.
[3] T. Moeller, *Ind. Eng. Chem. Anal.*, 1943, 15, 346.
[4] C. H. R. Gentry and L. C. Sherrington, *Analyst*, 1950, 75, 17.
[5] N. H. Furman, W. B. Mason and J. S. Pekola, *Analyt. Chem.*, 1949, 21, 1325.
[6] P. A. Clifford, *J. Assoc. Off. Agric. Chem.*, 1943, 26, 26.
[7] R. J. Lacoste, M. H. Earing and S. E. Wiberley, *Analyt. Chem.*, 1951, 23, 871.

Extraction of Gallium as 8-hydroxyquinolate. According to Moeller and Cohn,[1] gallium is completely extracted by four successive portions of 5–10 ml. chloroform solution of 8-hydroxyquinoline, each containing an amount of oxine equivalent to the expected total amount of gallium, from 25 ml. aqueous sulphate solution whose pH is adjusted by the addition of solid sodium acetate to just above pH 3·0. The extraction of the gallium is complete over the range pH 3·0–6·2, but the pH should be carefully controlled, as if this is not done other elements extractable at low pH will then also be removed. These include aluminium, bismuth, cobalt, copper, indium, iron-II, iron-III, nickel and tin-II. At pH values below 3·5 barium, cadmium, calcium, chromium, lead, magnesium, manganese, mercury-II, silver, strontium, tin-IV and zinc are not extracted appreciably.

Thallium is only extracted appreciably above pH 6·0, and maximum extraction (which is never complete) is over the narrow range pH 6·5–7·0, although thallium-III 8-hydroxyquinolate may be quantitatively precipitated from aqueous solution at pH 3·8 or over.

[1] T. Moeller and A. J. Cohn, *Analyt. Chem.*, 1950, **22**, 686.

Extraction of iron-III as acetonyl acetonate. Acetonyl acetone, $CH_3.CO.CH_2.CO.CH_3$, b.p. 139° C., d. 0·976, forms complexes with many metals. These are expressed with formulae derived from the enol form of the reagent, $CH_3.CO.CH : C(OH).CH_3$, and written in the general form

$$H_3C-C=O$$
$$|$$
$$HC \qquad \searrow M$$
$$\qquad \overline{n} \,,$$
$$|$$
$$H_3C-C-O \qquad /$$

where n is equal to the valency of the metal atom. Many of these complexes are quite stable, and are only moderately soluble in water, but are readily soluble in organic solvents, including acetonyl acetone itself.[1]

Iron-III may be extracted quantitatively from aqueous solution by using a solution of the reagent in carbon tetrachloride.[2] Since the iron complex is fairly readily hydrolysed by acid, and the iron is precipitated as hydroxide by alkali, the acid concentration should be such as to maintain the solution around neutrality.

Copper, manganese, titanium, vanadium and zirconium are also extracted. The method has been applied to the removal of iron before the determination of magnesium.

[1] L. E. Maley and D. P. Mellor, *Austral. J. Sci.*, 1949, **2A**, 92.
[2] E. Abrahamczik, *Mikrochem. Mikrochim. Acta*, 1947, **33**, 209.

Separation of hafnium from zirconium by thenoyltrifluoroacetone.
Zirconium and hafnium form complexes with thenoyltrifluoroacetone
(1-trifluoro-2-hydroxy-4-thienyl-*n*-but-2-en-4-one) which are represented by the formula:

When extraction is carried out by shaking a 2 N perchloric acid
solution of zirconium and hafnium with a benzene solution of the
reagent, a very much higher proportion of hafnium goes into the
organic phase, and although these two elements are so similar in
general chemical behaviour, a very good separation is obtained.[1]

Two extractions with 0·025 M solution of the reagent in benzene
of a solution containing zirconium and hafnium in the ratio of about
2 : 3 remove about one-third of the original hafnium, but only about
1 per cent of the zirconium. With a lower hafnium content separation
is even more satisfactory.

[1] E. H. Huffmann and L. J. Beaufait, *J. Amer. Chem. Soc.*, 1949, **71**, 3179.

CHAPTER III

INORGANIC PRECIPITANTS

A. MULTIPURPOSE

1. IODIC ACID

MERCURIC mercury may be determined by precipitation as iodate[1] in the presence of the alkali metals, aluminium, cadmium, calcium, chromium, cobalt, nickel, zinc and moderate amounts of barium, copper and strontium. Bismuth, iron, lead, silver titanium, zirconium and halide ions interfere. Preparation of a nitric acid solution of the sample precipitates antimony and tin. The determination may be completed gravimetrically or titrimetrically.

Precipitation of the iodate has been applied to the determination of lead. Earlier methods consisted in adding a measured amount of iodate, and determining the excess of iodate in the solution after removal of the precipitate. Gentry and Sherrington[2] have developed both a gravimetric and a titrimetric procedure based on direct determination.

Cold lead iodate solutions tend to supersaturate, and incomplete precipitation may result even after 20 hours standing. If, however, iodic acid solution is slowly added to the hot lead solution, and the whole is digested, the lead is quantitatively precipitated and the lead iodate is easily filtered. The optimum conditions for precipitation are in a solution which is 3 per cent in nitric acid. An excess of 0·75 g. iodic acid for each 100 ml. of solution is required.

Bismuth, mercury, silver, thorium, titanium and zirconium interfere. Iron-III is precipitated. Small amounts of barium or calcium do not interfere, but larger amounts prevent complete precipitation of the lead. Moderate amounts of aluminium, cadmium, magnesium, manganese, potassium, strontium, sodium and zinc have no effect.

The method has been applied successfully to the rapid determination of lead in glasses containing approximately 50 per cent silica,

30 per cent lead oxide, 14 per cent potassium oxide, 2 per cent sodium oxide, 2 per cent calcium oxide, 1 per cent alumina, small amounts of barium oxide and traces of iron, manganese and antimony oxides. Profitt and Chirnside[3] confirm the interference of iron and titanium, and suggest that if more than a trace of iron is present (as shown by the yellow colour of the precipitate) it is advisable to redissolve, and determine the lead as lead sulphate, since 0·07 per cent of Fe_2O_3 in the glass can introduce an error of 0·25 per cent in the lead oxide figure.

Rammelsberg[4] observed that alkali iodate solutions precipitated insoluble mercurous iodate from solutions of mercurous salts. This precipitate was investigated[5] as offering an alternative to the orthodox determination as mercurous chloride. Its solubility product was found to be $3·4 \times 10^{-13}$ at 15° C.

Quantitative precipitation is found to occur from warm moderately concentrated solutions, the precipitate forming in fine pearly needles. The gravimetric determination can be carried out with high accuracy.

By adding a known excess of alkali iodate, liberating iodine from the excess, and titrating this with standard sodium thiosulphate solution, an indirect titrimetric procedure is possible. Because of the multiplication effect, whereby one equivalent of iodate liberates six equivalents of iodine, according to the equation:

$$IO_3^- + 5I^- + 6H^+ \longrightarrow 3I_2 + 3H_2O$$

the accuracy of the mercury determination is increased still further.

The method of Bogdanov[6] in which barium is precipitated as the iodate, suspended in water, and determined iodometrically, has been modified by Guthrie[7] who obtained low results with the original method, using water as the wash liquid, and high results when saturated barium iodate was used as the wash solution. Guthrie also found that a gravimetric finish gave uncertain results after drying at 130° C., although the value obtained by titrating the precipitate agreed with its weight. Calcium, even in small amounts, caused serious errors.

Thorium may be determined iodometrically after precipitation as the iodate.[8] A complete separation from yttrium and the lanthanons is effected. Although lacking in absolute accuracy, the procedure gives results comparable with conventional methods, and is much more rapid. The lack of absolute accuracy arises from the difficulty of washing out adsorbed nitrate and iodate without losses arising from solubility and from hydrolysis.

Lanthanons are precipitated at acidities below 1·6 N to nitric acid, so that it is essential to use a fair excess of iodate to effect complete precipitation of thorium at the higher acidities necessary to prevent interference. Double precipitation is desirable if the amount of lanthanons and yttrium is large.

Immediate filtration of the precipitate is not possible, since it is too finely divided. However, prolonged heating renders the precipitate soluble only with difficulty in acids. It is therefore digested at room temperature, with frequent stirring.

Procedures

Gravimetric determination of mercuric mercury. Transfer the neutral solution containing the mercury compound to a 250-ml. beaker, dilute to 100 ml. with water, and add 1 ml. concentrated nitric acid. Boil, and add slowly, with vigorous stirring, a warm saturated solution containing 2 g. iodic acid. Leave on the hot plate for 5 minutes, stirring occasionally. Cool rapidly to room temperature. Filter the heavy white precipitate on an X4 sintered-glass filter crucible, and wash five times with a wash solution consisting of 1 per cent nitric acid containing 2 per cent iodic acid. Wash three times with small amounts of cold water, and dry for 1 hour at 140° C. Weigh as $Hg(IO_3)_2$.

$$mg. \ ppt. \times 0·3644 \equiv mg. \ Hg.$$

Titrimetric determination of mercuric mercury. Precipitate as in the gravimetric procedure, filter through a paper-pulp pad, and wash as before. Transfer pad and precipitate to a stoppered flask, washing the traces of precipitate from the funnel to the flask with 10 per cent potassium iodide solution. Add a further 3 g. solid potassium iodide, and dilute to 150 ml. Add 5 ml. concentrated hydrochloric acid, and titrate immediately with 0·1 N sodium thiosulphate solution, adding starch indicator just before the end-point.

$$1 \ ml. \ 0·1 \ N \ thiosulphate \equiv 1·672 \ mg. \ Hg.$$

Gravimetric determination of lead. Dilute the solution containing about 0·3 g. lead and 4 ml. nitric acid (sp. gr. 1·42) to 125 ml. Heat to boiling, and add slowly with stirring 25 ml. 6 per cent iodic acid solution. Stir for a further 2 minutes, and digest for 30 minutes at 60°–70° C. Cool, and filter through an X3 sintered-glass filter crucible. Wash the precipitate with 75 ml. of a solution which is 0·2 per cent in iodic acid and 1 per cent in nitric acid. Wash with three 2-ml. portions of cold water. Wash twice with acetone, dry for 1 hour at 140° C., and weigh as $Pb(IO_3)_2$.

$$mg. \; ppt. \times 0.3720 \equiv mg. \; Pb.$$

Titrimetric determination of lead. Proceed as in the gravimetric determination up to the washing with cold water. Then dissolve the precipitate in hot 10 per cent sodium hydroxide solution, and finally wash the crucible with hot 2 per cent sodium hydroxide solution, adding the washings to the bulk of the solution. Cool, and add 3–4 g. solid potassium iodide. Dilute to 150 ml., acidify with hydrochloric acid, and titrate with standard sodium thiosulphate solution. Alternatively, make to a known volume and titrate an aliquot.

$$1 \; ml. \; 0.1 \; N \; thiosulphate \equiv 1.727 \; mg. \; Pb.$$

Gravimetric determination of mercurous mercury. Render the mercurous solution neutral or weakly acid to nitric acid, adjust the volume to 60–80 ml., and heat. To the hot solution add dropwise, with constant stirring, an excess of a solution containing 3·5 g. potassium iodate per litre. Allow to cool for 1–1½ hours, filter through a porous-porcelain filter crucible, wash with water, 96 per cent ethanol and ether, and dry *in vacuo*. Weigh as $Hg_2(IO_3)_2$.

$$mg. \; ppt. \times 0.5342 \equiv mg. \; Hg.$$

Titrimetric determination of mercurous mercury. Render the solution neutral or slightly acid to nitric acid, heat and add a measured excess of a solution containing 3·525 g. potassium iodate per litre. The excess should be approximately 5–10 ml. per 10 ml. mercurous solution. Set aside for 1 hour for the precipitate to settle. Filter through a fine filter paper, and wash the vessel and precipitate with water. To the combined filtrate and washings add 0·5–1·0 g. potassium iodide, 0·5 ml. 2 N hydrochloric acid or sulphuric acid, and starch solution, and titrate the liberated iodine with 0·1 N sodium thiosulphate solution which has been standardised against iodine. Determine the strength of the potassium solution in the same way.

$$937 \cdot 4 \; (n - 0 \cdot 281 \; Tc) \equiv \text{mg. Hg,}$$

where n = weight of added potassium iodate in g.

T = titre of thiosulphate solution in g. iodine per ml.

c = ml. thiosulphate solution.

Titrimetric determination of barium. For the determination of barium in barium chloride dissolve 60–300 mg. of the salt in 20 ml. water and heat to boiling. Add 25 ml. of a solution of potassium iodate containing 32 g. per litre from a pipette through a dropping tube, and wash the latter with several small quantities of water. Boil for about 1 minute, and allow to cool to room temperature. Transfer to a 100-ml. graduated flask, make up to the mark with water, and mix well. Allow to stand for 1 hour, and filter through a dry paper. Neglect the first few ml. of the filtrate. Titrate 10-ml. portions. Alternatively, pipette 10 ml. portions of the supernatant liquid, taking care to wipe off any solid particles adhering to the outside of the pipette.

To the aliquot add 2 g. potassium iodide, 5 ml. 2 N sulphuric acid and 100 ml. water, and titrate the liberated iodine with 0·1 N sodium thiosulphate, which has been standardised against pure potassium iodate.

Determine the strength of the iodate solution by diluting 25 ml. to 100 ml. with water, and titrating a 10-ml. portion as above, using the same graduated flask and pipette.

The volume of the barium iodate may be allowed for by taking the density as 5·5, but the error caused if this is neglected only amounts to 1 part in 1000.

1 ml. 0·1 N thiosulphate ≡ 1·145 mg. Ba.

Titrimetric determination of thorium. Add 50 ml. concentrated nitric acid to 100 ml. of a solution containing about 175 mg. ThO_2. If cerium-IV is present, add a few drops 30 per cent hydrogen peroxide and remove the excess by boiling. Cool, and add 15 g. potassium iodate dissolved in 50 ml. concentrated nitric acid, and 30 ml. water. Allow to stand for 30 minutes with frequent stirring. Filter on a No. 50 Whatman filter paper, and wash with 20 ml. of a solution containing 2 g. potassium iodate and 25 ml. concentrated nitric acid in 225 ml. water.

Wash the precipitate back into the precipitation beaker with hot water. Heat to boiling, and add 30 ml. concentrated nitric acid, stirring until all the precipitate is dissolved. Cool to room temperature, and reprecipitate by adding a solution of 4 g. potassium iodate and 7 ml. concentrated nitric acid in 20 ml. water. Filter through the original paper, drain, and wash with 100 ml. ice-cold water in small portions.

Transfer the paper and precipitate to a 500-ml. conical flask and dissolve in 100 ml. 4 N sulphuric acid. Add 50 ml. water and 30–35 ml. 10 per cent potassium iodide solution. Titrate the liberated iodine immediately with 0·2 N sodium thiosulphate solution to the starch end-point.

$$1 \text{ ml. } 0·2 \text{ N thiosulphate} \equiv 1·934 \text{ mg. Th.}$$

Determination of thorium in Monazite sand. Add 10–20 g. Monazite sand concentrate (through 100 mesh) to concentrated sulphuric acid which has been evaporated to fuming. Boil until the sand is completely disintegrated, and cool to room temperature. Add crushed ice, and when the ice has melted filter through a Büchner funnel after adding paper pulp. Wash with water, and dilute the filtrate and washings to 500 ml. in a graduated flask.

Dilute an aliquot corresponding to 1–2 g. of the original sand to 100 ml., and proceed as described for the determination of thorium.

[1] C. H. R. Gentry and L. G. Sherrington, *Analyst*, 1945, **70**, 419.
[2] *Idem, ibid.*, 1946, **71**, 31.
[3] P. M. C. Proffitt and R. C. Chirnside, *ibid.*, 1947, **72**, 206.
[4] C. Rammelsberg, *Pogg. Ann.*, 1838, **44**, 570.
[5] G. Spacu and P. Spacu, *Z. anal. Chem.*, 1934, **96**, 30.
[6] K. A. Bogdanov, *Zavod. Lab.*, 1938, **1**, 793.
[7] F. C. Guthrie, *J. Soc. Chem. Ind.*, 1940, **59**, 98.
[8] T. Moeller and N. D. Fritz, *Ind. Eng. Chem. Anal.*, 1948, **20**, 1055.

2. PERIODIC ACID AND PERIODATES

Periodic acid and its salts have found some application as analytical reagents. Their use in the colorimetric determination of manganese, by oxidising the manganous ion to permanganate has been well known for some considerable time, and is not appropriate for discussion here, where we are more concerned with determinations which may be completed gravimetrically or titrimetrically.

Mercury is usually weighed as the metal, or as mercuric sulphide. The latter weighing form is not very satisfactory in the presence of cadmium, copper or zinc, because of contamination of the precipitate. Precipitation as mercuric *para*periodate, $Hg_5(IO_6)_2$, offers a rapid and accurate method.

Iron interferes and must be removed. Complexing of iron as the $FeF_6^{\equiv}$ ion is not possible, since the precipitation of mercury as the *para*periodate is not then complete. Halides interfere owing to complex formation. If they are present the mercury should first be precipitated as the metal or the sulphide. Moderate amounts of aluminium, cadmium, calcium, copper, magnesium, nickel and zinc are without effect.

The maximum permissible acidity at which the precipitation may be effected is 0·15 N nitric acid or 0·1 N sulphuric acid. The method may be completed gravimetrically or titrimetrically by one of two alternative procedures.[1]

The first of these depends on the following overall reaction:

$$Hg_5(IO_6)_2 + 34KI + 24HCl \longrightarrow$$
$$5K_2HgI_4 + 8I_2 + 24KCl + 12H_2O.$$

The liberated iodine is then titrated by standard thiosulphate solution.

In the second method, excess of standard arsenite solution is added to the precipitate. Mercuric *para*periodate reacts as follows with arsenic-III:

$$Hg_5(IO_3)_2 + 6H_3AsO_3 + 12HCl \longrightarrow$$
$$5HgCl_2 + 2ICl + 6H_3AsO_4 + 6H_2O,$$

and standard iodate is used to determine the excess of arsenite.

Potassium may be precipitated quantitatively as the periodate, KIO_4. The method[2] is rapid and accurate. It may be completed gravimetrically, a favourable conversion factor resulting from the low potassium content. A titrimetric finish is also possible, the precipitate being dissolved in a boric acid-borax buffer, potassium iodide being added, and the liberated iodine being titrated with sodium arsenite. The periodate is reduced to iodate in this reaction, according to the following equation:

$$IO_4^- + 2I^- \longrightarrow IO_3^- + I_2$$

The reaction has the advantage that iodate, if present, does not interfere, as it would if the periodate were reduced to iodide. In the latter process, however, the equivalent would be much smaller.

5

Using this reagent potassium may be separated from aluminium, calcium, cobalt, lithium, magnesium, nickel, sodium and zinc. The procedure may be applied to the mixed chlorides obtained by the Lawrence-Smith method, provided that they are first converted to nitrates by evaporation with concentrated nitric acid. As little as 0·4 mg. potassium may be separated from seventy times as much sodium.

Ammonium, caesium, chromium, iron, manganese and rubidium interfere. When calcium and sulphate are present together low results are obtained, probably because a double potassium calcium sulphate is formed. Small amounts of sulphuric, nitric, phosphoric and boric acids may be present, but with larger amounts the precipitate is rendered gelatinous and becomes difficult to filter. The maximum permissible amounts appear to be 0·5 ml. concentrated nitric acid, 5 drops concentrated sulphuric acid, 1 ml. concentrated phosphoric acid or 200 mg. boric acid.

The gravimetric finish cannot be employed in the presence of sulphate because the precipitate is contaminated with sodium sulphate, but in such conditions the titrimetric finish gives satisfactory results. Chloride must be absent since it causes reduction of the periodate ion.

Precipitation is effected in a small bulk of solution, and is completed by the addition of a mixture of equal parts of aldehyde-free ethanol and anhydrous ether. The solution is maintained at 0°C. and stirred for 30 minutes.

In dilute acid solution lead is quantitatively precipitated as triplumbic *para*periodate, $Pb_3H_4(IO_6)_2$, and the determination may be completed gravimetrically or titrimetrically.[3] In the presence of concentrated hydrochloric acid the precipitate reacts with arsenious acid according to the equation:

$$Pb_3H_4(IO_6)_2 + 8HCl + 6H_3AsO_3 \longrightarrow$$
$$3PbCl_2 + 2ICl + 6H_3AsO_4 + 6H_2O.$$

If arsenious acid is present in excess all the chlorine reacts with it to form arsenic acid. The unconsumed arsenious acid may then be titrated with potassium iodide.

The maximum permissible acidity is 0·025 N to nitric acid. However, for very small amounts of lead precipitation does not occur until 1 ml. precipitant has been added, and it is then necessary to decrease the acidity to 0·006 N, as otherwise the precipitate will not have the theoretical composition. After precipitation the solution must be cooled to 0° C. and stirred for 30 minutes, since supersaturation is pronounced.

The possibility of separating lead by this method from other metals is somewhat limited, since most metals form insoluble periodates at the low acid concentration necessary. However, the lead may be separated satisfactorily from moderate amounts of aluminium, cadmium, calcium, copper, magnesium, nickel and zinc. A double precipitation may be advantageous for larger amounts.

Lithium is precipitated quantitatively as a complex periodate by a strongly alkaline solution of potassium periodate.[4] By proper adjustment of the alkalinity the other alkali metals are not precipitated. Metals other than those of the alkali group must be absent. Ammonium should be removed by boiling with a slight excess of concentrated potassium hydroxide solution, since it increases the solubility of the precipitate.

There is no integral stoicheiometric relationship between the proportions of lithium and iodine, since the precipitate contains somewhat less lithium than is demanded by the formula Li_5IO_6. The precipitate probably consists of a mixture of lithium periodates. Nevertheless, under strictly controlled conditions the composition of the precipitate is constant, and the lithium content may be determined accurately by measurement of the periodate or iodine content.

The composition of the precipitate also varies according to the drying time, therefore a gravimetric finish is not satisfactory. Similarly, variations occur according to the temperature of precipitation.

When sodium is present in amounts greater than 20 mg. it interferes if precipitation is effected at high temperatures. However, as much as 150 mg. is without effect when precipitation is carried out in the cold. Apparently a periodate containing sodium separates at the higher temperatures. When there is doubt as to the amount of sodium present precipitation should always be effected at room temperature. If, however, it is known that not more than 20 mg. is present, the higher temperature is to be preferred for precipitation since the precipitate obtained is easier to filter and wash.

Although a titrimetric finish overcomes the variations due to drying, it will be obvious that the solution must be standardised under the conditions of the actual test, *i.e.*, according to whether the unknown is precipitated in the cold or from hot solution.

The range of the method extends up to 50 gm. lithium, and as little as 0·1 mg. lithium may be determined by restricting the volume of reagent and test solution. Two titrimetric procedures have been described. In the first, solution of the precipitate in concentrated sulphuric acid is followed by addition of potassium iodide and titration of the liberated iodine by standard sodium thiosulphate solution. Alternatively potassium iodide may be added to a buffered solution, the titration being made with sodium arsenite. The first method is more suitable for small amounts of lithium because of the favourable ratio between the volume of standard solution and the amount of lithium present. For large amounts of lithium the second method is preferable.

Lithium may also be determined[5] by precipitating as lithium potassium ferric periodate, $LiKFeIO_6$, and, after filtering and washing the precipitate, titrating this with a standard solution of thiosulphate or arsenite. Analysis of the precipitate obtained in a 0·8–1·0 N solution of potassium hydroxide shows that its formula is that already stated. When a more dilute solution of alkali is used the precipitate has a different composition; in 0·2 N alkali the composition corresponds to $LiK_3Fe_2(IO_6)_2$. This precipitate hydrolyses and is more difficult to filter. The alkali concentration should not be higher than 1·1 N.

Up to 50 per cent excess of reagent may be added without influencing the results. Above this amount positive errors are obtained which increase with excess of the reagent. The standard procedure, which permits up to a 50 per cent excess of reagent, will determine as little as 0·027 mg. lithium. However, for amounts of lithium less than this and down to 0·014 mg. a 100 per cent excess of reagent must be used to ensure complete precipitation. At these low concentrations positive errors are apparently not obtained or may compensate for solubility losses.

When sodium is present in amounts greater than 0·1–0·15 mg. per ml., high results are obtained. Sodium may, however, be separated readily by precipitating with an alcoholic solution of hydrochloric acid. The amount of sodium remaining in the filtrate is not sufficient to interfere. Potassium, rubidium and caesium do not interfere.

The volume of the solution should be less than the anticipated volume of precipitant, and the solution should be 0·8–1·0 N in potassium hydroxide. Solutions which are too dilute should therefore be concentrated. If 0·1 mg. or less of lithium is present the solution should be taken to dryness and the residue dissolved in 0·5–1·0 ml. 0·8–1·0 N solution of potassium hydroxide.

The precipitate is dissolved and titrated either with sodium thiosulphate or with sodium arsenite. Details concerning the use of the former titrant are rather obscure, but iron-III takes part in the reduction, so that 0·1 mg.-atom of iron-III together with 0·1 mg.-atom of iodine-VII liberate $9 \times 0·1$ mg.-atom of elementary iodine. This then consumes 9 ml. 0·1 N sodium thiosulphate solution.

For the arsenite titration more precise details are available. Under the conditions employed, only the iodine corresponding to periodic acid is liberated, according to the equation:

$$IO_4^- + 7I^- + 8H^+ \longrightarrow 4H_2O + 4I_2.$$

The reaction

$$2Fe^{+++} + 2I^- \rightleftharpoons 2Fe^{++} + I_2$$

proceeds from right to left after adding the potassium bicarbonate, since ferric hydroxide precipitates out.

The components of the precipitating reagent should be practically free from sodium. Potassium hydroxide is likely to contain sodium, but the amount should not be more than 0·08 per cent.

Reagents

Pure ethanol-ethyl acetate mixture. Reflux a quantity of 95 per cent aldehyde-free ethanol for 2–3 hours in the presence of 0·5 g. sodium hydroxide and 2·5 g. silver nitrate per 1000 ml. Distil, and mix with an equal volume of anhydrous ethyl acetate. The ethyl acetate increases the resistance of the ethanol to oxidation.

Precipitating Reagent I (Rogers and Caley). Dissolve 24 g. potassium hydroxide in 100 ml. water. When nearly cool add 10 g. potassium *meta*periodate. Keep the solution in a dark paraffin wax-lined bottle. It is stable for at least 1 month.

Precipitating Reagent II (Pročke and Šlouf). Dissolve 2·3 g. finely powdered potassium periodate in 10 ml. 2 N potassium hydroxide solution. Dilute to about 40 ml. with water, and add 2·0 ml. molar ferric chloride (270·3 g. $FeCl_3.6H_2O$ in 1 litre). Add a further 40 ml. 2 N potassium hydroxide solution, and when the ferric periodate is completely dissolved dilute to 100 ml. with water. This is only stable for one week when stored in glass.

Sodium arsenite solution, 0·1 N. Dissolve 4·945 g. arsenious oxide in a solution containing 10 g. sodium bicarbonate. Warm to about 80° C., saturate with carbon dioxide, cool, and dilute to 1000 ml. Standardise against pure potassium or sodium periodate, since it may differ from the theoretical by as much as 0·0012 N.

Ethanolic hydrogen chloride. Pass dry hydrogen chloride from a suitable generator through a washing bottle containing ethanol and then through a further washing bottle containing 0·1 N potassium hydroxide coloured with methyl orange. When the latter turns red the saturation of the ethanol is complete.

Procedures

Gravimetric determination of mercuric mercury. Adjust the acidity of the solution so that it is not more than 0·15 N to nitric acid or 0·1 N to sulphuric acid. The final bulk should be about 150 ml. Heat the solution to boiling, and add slowly, with stirring, 2·0 g. sodium or potassium periodate dissolved in 50 ml. water. Cool the solution, and filter on an X4 sintered-glass filter crucible. Wash the precipitate with warm water, dry at 100° C. for 2–3 hours, and weigh as $Hg_5(IO_6)_2$.

$$mg. \ ppt. \times 0·6922 \equiv mg. \ Hg.$$

Titrimetric determination of mercuric mercury, Method I. Attach the crucible containing the precipitate to a 150-ml. filter flask, and treat the precipitate with 2·0 g. potassium iodide and 10–15 ml. water. Stir until solution is complete and draw the contents into the filter flask. Wash the crucible with small amounts of water, drawing the washings into the flask. Add 10 ml. 2 N hydrochloric acid, and titrate the liberated iodine with 0·1 N sodium thiosulphate solution, using starch indicator.

$$1 \ ml. \ 0·1 \ N \ sodium \ thiosulphate \equiv 6·269 \ mg. \ Hg.$$

Titrimetric determination of mercuric mercury, Method II. Treat the precipitate with a measured excess of standard sodium arsenite solution, and then add concentrated hydrochloric acid until solution is complete. About 35 ml. should be used if the final volume antici-pated at the end of the titration is about 100 ml. Transfer the solution to a 150-ml. stoppered flask. Titrate the excess sodium arsenite with 0·1 N potassium iodate to a light brown colour, add 4–5 ml. chloroform, and continue the titration until the purple colour disappears from the chloroform layer. The solution should now contain 28–45 ml. of concentrated hydrochloric acid per 100 ml. solution.

$$1 \text{ ml. } 0\text{·}1 \text{ N sodium arsenite} \equiv 8\text{·}359 \text{ mg. Hg.}$$

Gravimetric determination of potassium. Dissolve the sample (0·1–0·2 g.) in 4–5 ml. water in a 150-ml. beaker. If chloride is present add 10 ml. concentrated nitric acid and evaporate to dryness before dissolving in water. Add 1·0 g. periodic acid (free from iodate) dissolved in 3 ml. water immediately before required, stir, and allow 3–4 minutes for the potassium periodate to precipitate. Add 90 ml. ethyl acetate-ethanol mixture, and stand the whole in an ice-bath for 30 minutes, stirring mechanically. If the periodic acid is added in the presence of the organic solvents the precipitate will be gelatinous instead of crystalline.

Filter through a Gooch crucible and wash with anhydrous ethyl acetate cooled to 0°C. Dry in an oven for 10 minutes at 105°C. and weigh as KIO_4. When the amount of potassium is very small the time allowed for complete precipitation should be extended to 1 to 1½ hours.

$$\text{mg. ppt.} \times 0\text{·}1700 \equiv \text{mg. K.}$$

Titrimetric determination of potassium. Precipitate the potassium periodate as described above, place the crucible in a 250-ml. beaker, and add 125 ml. of a solution containing 5 g. boric acid and 5 g. sodium *meta*borate. The potassium periodate dissolves readily in this solution, which has a pH of about 7·5.

When solution is complete add 3·0 g. potassium iodide, and without removing the crucible titrate the liberated iodine immediately with 0·1 N sodium arsenite solution.

$$1 \text{ ml. } 0\text{·}1 \text{ N sodium arsenite solution} \equiv 2\text{·}585 \text{ mg. K.}$$

Gravimetric determination of lead. Adjust the solution to be 0·025 N to nitric acid and to occupy a volume of about 200 ml. Heat to boiling, and add slowly 2 g. sodium periodate dissolved in 50 ml. water. Cool the solution in ice-water and stir for 30 minutes. Filter on a porous-porcelain filter crucible, wash with ice-water, and dry for 2 hours at 110°C. Weigh as $Pb_3H_4(IO_6)_2$.

$$\text{mg. ppt.} \times 0\text{·}5802 \equiv \text{mg. Pb.}$$

Titrimetric determination of lead. Filter the precipitate on a Gooch crucible, the bottom of which is covered with a filter-paper disc. Wash thoroughly, and transfer to a 150-ml. conical flask. Add a measured excess of standard sodium arsenite solution. Slowly add concentrated hydrochloric acid to the cold solution, whilst agitating the flask until all the precipitate has dissolved. About 35–40 ml. acid will be required. Titrate with 0·1 N potassium iodate to a light brown colour. Add a small amount of chloroform and complete the titration as described for mercury.

1 ml. 0·1 N sodium arsenite≡5·18 mg. Pb.

Titrimetric determination of lithium, Method I (Rogers and Caley). Remove all metals other than the alkali metals from the chloride, nitrate, perchlorate or sulphate solution containing not more than 50 mg. lithium. If ammonium is present in an amount of more than a few mg., remove by boiling with a slight excess of potassium hydroxide solution. Reduce the volume to 2 ml., finally transferring the solution to a 50-ml. beaker. Immerse the lower half of the beaker in a bath maintained at 60°–70° C. and add 2 ml. precipitating reagent I dropwise, with constant swirling, at a rate not exceeding one drop every 5 seconds. If a heavy precipitate forms, add a further 3 ml. precipitant in exactly the same way. If the precipitate is very heavy, as shown by a solution which does not flow freely, add yet a further 3–5 ml. Digest at 60°–70° C. for 20 minutes, and filter through a Gooch crucible fitted with a moderately thick asbestos pad. Wash with four successive 2-ml. portions of 3–5 N potassium hydroxide solution added slowly from a pipette.

If more than 20 mg. sodium were present, precipitate, filter and wash in exactly the same way, but at room temperature.

Transfer the pad and precipitate to a 250-ml. beaker with the aid of distilled water, and add 5 ml. N sulphuric acid to ensure complete solution of the precipitate. Add further N sulphuric acid if necessary to bring the volume to 10 ml. for every 25 ml. of the original solution. Add 2 g. potassium iodide and titrate with 0·1 N sodium thiosulphate solution using starch indicator.

(Precipitated hot) 1 ml. 0·1 N sodium thiosulphate≡0·407 mg. Li.

Titrimetric determination of lithium, Method II. After treatment as described in Method I and solution of the precipitate, add an excess of borax or sodium bicarbonate as a buffer before adding the potassium iodide. Titrate with standard sodium arsenite solution. If the amount of lithium is relatively low, add the potassium iodide before the buffering substance, and then titrate.

Standardise the titrating solution, in this and the previous determination, against a known amount of lithium, using exactly the same procedure as used in the determination.

(Precipitated cold) 1 ml. 0·1 N sodium arsenite ≡
1·581 mg. Li.

(Precipitated hot) 1 ml. 0·1 N sodium arsenite ≡
1·638 mg. Li.

Determination of lithium as triple periodate in absence of sodium (*Pročke and Šlouf*). Place the solution, contained in a beaker, or, better still, in a platinum vessel, on a boiling water bath. Add precipitating reagent II gradually whilst stirring and maintaining the solution above 90° C. until precipitation is complete and the supernatant liquid is distinctly yellow, denoting a slight excess of precipitant. After about 5 minutes set the vessel aside to cool and then filter through an X4 sintered-glass filter crucible. Wash two or three times with 2–5 ml. lots of about 0·25 N potassium hydroxide solution. Disperse the precipitate in water, add potassium iodide and a sufficient small excess of hydrochloric acid to dissolve the precipitate. After a time titrate the solution with standard sodium thiosulphate solution.

Alternatively dissolve the precipitate in the minimum amount of dilute sulphuric acid, and add a solution of potassium hydroxide dropwise until a permanent turbidity results. Just remove this by dropwise addition of dilute sulphuric acid. Add excess of potassium iodide, and, after a time, an excess of potassium bicarbonate. Titrate with standard sodium arsenite solution.

1 ml. 0·1 N sodium thiosulphate ≡ 0·0771 mg. Li.
1 ml. 0·1 N sodium arsenite ≡ 0·0868 mg. Li.

Determination of lithium as triple periodate in the presence of sodium. Evaporate the solution of chlorides or sulphates to about 1 ml. in a conical flask. Add 10–15 ml. ethanol saturated with hydrogen chloride, and shake occasionally. After 1 hour filter on an X4 sintered-glass filter crucible, and wash three times with 2 ml. ethanol saturated with hydrogen chloride. Evaporate the filtrate to dryness, dissolve the residue in at least 1 ml. 0·8–1·0 N potassium hydroxide solution, and proceed as before.

[1] H. H. Willard and J. J. Thompson, *Ind. Eng. Chem. Anal.*, 1931, **3**, 398.
[2] H. H. Willard and A. J. Boyle, *ibid.*, 1941, **13**, 137.
[3] H. H. Willard and J. J. Thompson, *ibid.*, 1934, **6**, 425.
[4] L. B. Rogers and E. R. Caley, *ibid.*, 1943, **15**, 209.
[5] O. Pročke and A. Šlouf, *Coll. Czech. Chem. Comm.*, 1939, **11**, 276.

3. POTASSIUM COBALTICYANIDE

Potassium cobalticyanide, $K_3[Co(CN)_6]$, which resembles potassium ferricyanide in most of its reactions, but has the advantage of being much more stable, gives precipitates with many metallic ions. Certain of its reactions can be used, either with a view to determination of the precipitated ion, or with a view to subsequent determination of an ion not precipitated after removal of an interfering element as the cobalticyanide.

Under proper conditions bismuth, cadmium, cobalt, copper-I, copper-II, iron-II, manganese-II, nickel, silver, vanadium-IV and zinc can be completely precipitated. The following do not give precipitates in hydrochloric, nitric or sulphuric acid solutions: aluminium, ammonium, antimony, arsenic, barium, beryllium, calcium, cerium-III, cerium-IV, chromium-III, chromium-VI, iron-III, lead, lithium, magnesium, molybdenum-VI, potassium, selenium, sodium, strontium, tellurium, tin-II, tin-IV, titanium-IV, uranium, vanadium-V and zirconium.

Analytical procedures have been worked out for the following determinations:

A. *Separation of an ion which is subsequently determined.*
  (i) Determination of nickel in iron.[1]
  (ii) Determination of nickel and cobalt in iron.[1]
  (iii) Determination of cadmium in lead-tin-cadmium alloys.[2]
  (iv) Determination of cadmium in lead-antimony-cadmium alloys.[2]
  (v) Determination of silver in lead.[2]
  (vi) Determination of manganese in citrate solutions.[2]
  (vii) Determination of iron in aluminium.[2]
  (viii) Determination of vanadium in aluminium.[2, 3]

B. *Removal of an interfering ion.*

(i) Separation of copper from selenium and tellurium, with subsequent determination of the selenium and tellurium.[4]

Procedures

Determination of nickel in iron. Oxidise the iron in the solution containing about 0·01–0·02 g. nickel with nitric acid. Neutralise approximately with sodium hydroxide solution, and add 50 ml. 10 per cent aqueous potassium nitrate solution. Dilute to about 100 ml. with water, and add 10 ml. syrupy phosphoric acid and 5 ml. concentrated nitric acid. Cool thoroughly, and add excess of a saturated solution of potassium cobalticyanide in 5 per cent nitric acid which has been prepared hot, cooled and filtered. Add 20 ml. 1 per cent gum arabic solution and a little amyl alcohol to aid filtration, followed by paper pulp. Set aside for 45–60 minutes, filter off the green precipitate on a tight paper-pulp pad, wash eight times with hot 5 per cent potassium nitrate solution, and transfer the pulp pad and precipitate to the original beaker, washing it in with 100 ml. 5 per cent sodium hydroxide solution. Stir well, boil for a few seconds, cool and add 2–3 ml. 6 per cent hydrogen peroxide. Stir occasionally for 10 minutes, when the nickel will have been transformed to hydroxide. Filter immediately through paper pulp, and wash with 5 per cent sodium chloride solution.

Dissolve the precipitate through the filter into a clean flask by a boiling mixture of 5 N hydrochloric acid and saturated sulphur dioxide solution, followed by thorough washing of the filter with water. Boil down the solution, which contains nickel chloride with only a trace of iron, to about 10 ml., and add excess of ammonia to precipitate the iron. Filter, washing the precipitate with hot water, and join the washings to the filtrate. Redissolve the precipitate in hydrochloric acid, reprecipitate with ammonia, and filter into the original filtrate, together with the hot water washings. Take the total filtrate down to a volume of about 50 ml.

If cobalt is absent, cool, add 10 ml. 7 N ammonia and 10 ml. 4 per cent potassium iodide solution, dilute to 200–300 ml. with cold water, and add standard potassium cyanide solution from a burette until the precipitate has just dissolved. Read the volume of potassium cyanide solution, V, and add an excess to bring this up to $(1·3\ V + 5)$ ml. Add 10 ml. 10 per cent sodium carbonate solution, filter off any slight precipitate, and wash the precipitate with cold water, joining the washings to the filtrate. Pass a rapid stream of air through the solution for 6 minutes, add 10 ml. 7 N ammónia and 25 ml. 20 per cent ammonium chloride solution, and titrate the clear liquid just to opalescence with a standard solution of silver nitrate containing 5·792 g. per l. Add potassium cyanide solution dropwise from the burette to the point at which the opalescence just disappears.

Standardise the potassium cyanide solution against the silver nitrate solution by titrating to the same end-point, and convert the total volume of cyanide used in the nickel titration into terms of silver nitrate solution. The difference between the total equivalent of cyanide solution used and the silver nitrate used gives the amount of nickel present in terms of the standard silver solution.

1 ml. silver nitrate (5·792 g. per l.)≡1·000 mg. Ni.

If cobalt is present in addition to iron and nickel, the procedure is slightly different from the point at which removal of iron has been completed. As before, take down the total filtrate to a volume of about 50 ml. Make the solution slightly ammoniacal, add 5 ml. 10 per cent *sodium* cyanide solution and a few drops 20-volume hydrogen peroxide, and boil for 5 minutes to convert the cobalt to cobalticyanide. Add 20 ml. 20 per cent ammonium chloride solution, and boil for a further 15 minutes. Cool slightly and add 10 ml. 7 N ammonia and 10 ml. 6 per cent hydrogen peroxide, again boiling very gently for 10 minutes with the flask covered. Cool, and carry out the titration as described for nickel in the absence of cobalt.

Determination of nickel and cobalt in iron. Use two equal aliquots of the solution, and treat one exactly as directed for the determination of nickel, carrying out the full procedure for nickel when cobalt is present.

Treat the second aliquot in the same way up to the point where the nickel is converted to chloride and still contaminated with a trace of iron. The cobalt, which as cobalticyanide precipitate was rose-pink, will also have been converted to chloride. Evaporate the solution of mixed chlorides almost to dryness, and precipitate the nickel and cobalt with the minimum amount of sodium hydroxide solution. Redissolve the hydroxides in the minimum amount of 8 N nitric acid. Add 20 ml. saturated borax solution to reprecipitate, inducing precipitation, if necessary, by dropwise addition of dilute sodium hydroxide solution, shaking after each drop. Add 10 ml.

4 per cent potassium iodide solution, and carry out a titration with standard potassium cyanide solution in the same way as described for the nickel determination. Once again convert the total volume of cyanide used into terms of the standard silver nitrate solution. Subtract the volume of silver nitrate solution used from this, and also the silver nitrate equivalent of the nickel, as already determined on the first aliquot of solution. The final figure gives the cobalt content of the aliquot in terms of the standard silver solution.

1 ml. silver nitrate (5·792 g. per 1.)≡0·803 mg. Co.

Determination of cadmium in lead-tin-cadmium alloys. Dissolve 5 g. alloy in 30 ml. 8 N nitric acid. Evaporate to dryness. Boil the residue for several minutes with 100 ml. of an aqueous solution containing 5 g. potassium nitrate and 5 ml. concentrated nitric acid. Filter through a close paper-pulp pad, and wash with further portions of the potassium nitrate-nitric acid solution, adding the washings to the filtrate. Add 15 ml. concentrated nitric acid for each 100 ml. filtrate, and follow this with 20 ml. 10 per cent aqueous potassium cobalticyanide solution. Stir vigorously and allow to settle for 30 minutes.

Filter off the white flocculent precipitate on a close paper-pulp pad, and wash several times with 1 per cent potassium cobalticyanide solution in 20 per cent nitric acid. Rinse the beaker with 10 ml. 10 per cent sodium hydroxide solution, passing this through the filter, and then wash the filter with a further 20 ml. 10 per cent sodium hydroxide solution. Finally wash the precipitate, which is now completely converted to cadmium hydroxide, several times with water. Dissolve the hydroxide by passing 50 ml. N sulphuric acid through the filter and collecting it in a small beaker. Wash the filter with 50 ml. water in small portions, and add the washings to the main bulk of solution. Precipitate the cadmium as sulphide with sulphuretted hydrogen in the cold, ignite to the sulphate at low temperature, treat with a little sulphuric acid, reignite at 400°–500° C. and weigh as $CdSO_4$.

mg. ppt. × 0·5392≡mg. Cd.

Determination of cadmium in lead-antimony-cadmium alloys. To 30 ml. 8 N nitric acid add 20 ml. 50 per cent aqueous citric acid solution. Use this to dissolve 5 g. of the alloy, and evaporate the solution almost completely to dryness. Take up in 150 ml. 3 N nitric acid, and make up to 200 ml. with water. To the cooled solution add 20 ml. 10 per cent aqueous potassium cobalticyanide solution, and complete the determination of cadmium as already described.

Determination of silver in lead. Dissolve 50 g. lead in 200 ml. 8 N nitric acid. Boil to remove nitrous fumes, and add 400 ml. hot water. Cool, and add 10 ml. 10 per cent aqueous potassium cobalticyanide. After allowing to stand for 2 hours filter off the precipitate, which is flocculent and settles readily, on a close paper-pulp pad. Wash the precipitate with 1 per cent potassium cobalticyanide in N nitric acid. Dissolve the precipitate from the pad by 40 ml. 4 N ammonia added in small portions, and wash through several times with water. If the filtrate is not completely clear digest for half an hour on the steam bath, and filter. Neutralise with nitric acid, add 10 ml. 7 N ammonia and 0·2 g. solid potassium iodide. Dilute to 100 ml. with water, run in a measured excess of standard potassium cyanide solution, and back titrate with standard silver solution until a permanent turbidity, seen against a dark background, is obtained.

1 ml. 0·1 N potassium cyanide≡10·79 mg. Ag.

Determination of manganese in citrate solution. This determination is a difficult one if carried out by the methods usually recommended— complete destruction of the citrate, or precipitation of the manganese as sulphide. A cobalticyanide precipitation is to be preferred. To 100 ml. of the solution, which should be about 10 N to sulphuric acid, add 5 ml. glacial acetic acid followed by 10 ml. 10 per cent aqueous potassium cobalticyanide solution. Set aside for 2 hours and filter off the fine white precipitate through paper pulp. Wash with 10 N sulphuric acid containing 5 per cent of glacial acetic acid and 1 per cent of potassium cobalticyanide. Transfer pad and precipitate to the original vessel, add 90 ml. 10 per cent sodium hydroxide solution and 10 ml. 6 per cent hydrogen peroxide. Stir thoroughly to break up the pad, and allow to stand for 10 minutes, stirring frequently. Do not warm the solution, as this will cause loss of manganese due to reduction of the dioxide by paper pulp.

Filter through paper pulp which has first been boiled for half an hour with 0·5 per cent hydrogen peroxide. Wash with cold water. Transfer the pulp and precipitate to the original beaker. Add 30 ml. 8 N nitric acid, 30 ml. water and 5 ml. saturated aqueous sulphur dioxide. Warm to dissolve the manganese dioxide, and filter, retaining the filtrate. Wash the pad with water, and combine the washings with the filtrate. Boil to reduce the volume of solution to about 30 ml. Cool the solution to room temperature, and add 0·5–1 g. sodium bismuthate. Stir for a few minutes, and if no bismuthate remains undissolved, add further small amounts until excess is present. Add 50 ml. cold 0·5 N nitric acid, and filter through asbestos or sintered glass, washing the residue well with dilute nitric acid.

Add 3 ml. syrupy phosphoric acid and a measured slight excess of standard ferrous ammonium sulphate solution. Back titrate with standard permanganate to the first faint permanent pink. Carry out a blank determination using approximately the same amount of ferrous ammonium sulphate solution, and use the figure obtained to standardise the ferrous solution, thus avoiding an indicator correction.

1 ml. 0·1 N ferrous ammonium sulphate $\equiv$ 1·099 mg. Mn.

Determination of iron in aluminium. To 50 ml. neutral solution containing 5 g. aluminium add 2 g. sodium chloride and 50 ml. 10 N hydrochloric acid. Add 10 ml. 10 per cent aqueous potassium cobalticyanide solution followed by 0·5–1·0 g. solid potassium iodide. On stirring thoroughly the iodide reduces the iron to iron-II and ferrous cobalticyanide is precipitated, flocculation being aided by the liberated iodine.

Add 10 ml. saturated aqueous sulphur dioxide solution, and allow to stand for 1 hour. Filter off the precipitate through paper pulp, and wash with 5 N hydrochloric acid containing 2 per cent sodium chloride and 1 per cent potassium cobalticyanide. Transfer the pulp and precipitate to the original vessel, and add 100 ml. 20 per cent ammonium sulphate solution, 50 ml. 3 N ammonia, and 10 ml. 6 per cent hydrogen peroxide. Boil, and filter off the paper pulp and ferric hydroxide. Wash with 5 per cent ammonium sulphate solution, ignite and determine the iron either titrimetrically or gravimetrically.

Determination of vanadium in aluminium. The vanadium must be present as vanadium-IV, and hydrochloric acid must be absent. Make the solution 10 N in sulphuric acid, and add aqueous potassium cobalticyanide solution. Stir in a little paper pulp and shake for a few minutes to flocculate the gelatinous sky-blue precipitate. Allow to stand for 30 minutes, and filter through paper pulp. Wash the precipitate with 10 N sulphuric acid containing 1 per cent potassium cobalticyanide. Dissolve the precipitate through the filter into a 600-ml. beaker with a mixture of 50 ml. 7 N ammonia and 10 ml. 10 per cent aqueous potassium cyanide solution, washing the pad several times with cold water. Pour 30 ml. boiling 5 N sulphuric acid through the filter to remove traces of vanadyl hydroxide, and finally wash the filter with cold water. Evaporate the filtrate to a volume of about 250 ml. Cool, and make slightly acid with sulphuric acid and then slightly alkaline with sodium bicarbonate solution. Add 10 ml. 10 per cent aqueous potassium cyanide solution, 2–3 g. sodium hydrosulphite and 50 ml. saturated aqueous borax solution. Boil for 5 minutes, add 20 ml. 20 per cent sodium hydroxide solution, and boil for a further 5 minutes.

Filter the solution hot through a close paper-pulp pad, wash the precipitate with 5 per cent aqueous sodium sulphate solution, and burn off the filter in a platinum dish at as low a temperature as possible. Fuse the residue in the platinum dish with a little fusion mixture and dissolve in water.

Adjust the solution with 5 N hydrochloric acid so that it is only slightly alkaline, and transfer to an Erlenmeyer flask fitted with a three-holed rubber stopper. In the stopper are fitted a carbon dioxide lead-in tube coming to within half an inch of the surface of the liquid, a small tap funnel and a removable glass plug. Add 5 ml. 10 per cent freshly prepared potassium ferrocyanide solution, and 50 ml. 5 N hydrochloric acid. Place the stopper in position, and sweep out for 5 minutes with a rapid stream of carbon dioxide. Remove the glass plug, and add through the tap funnel 20 ml. 4 per cent potassium iodide solution and 1 ml. 10 per cent zinc sulphate solution, and rinse these in with 30 ml. water without allowing air to enter through the tap funnel during the additions. Reinsert the glass plug, shake the flask, and set aside for 5 minutes. The vanadium-V is converted to vanadium-IV by the ferrocyanide, with production of an equivalent amount of ferricyanide. The ferricyanide, in the presence of zinc ions as catalyst, oxidises the potassium iodide quantitatively to iodine.

After the necessary time has elapsed replace the glass plug by the tip of a burette filled with 0·01 N sodium thiosulphate solution. Titrate until the *solution* is almost colourless (bearing in mind that the precipitated vanadium ferrocyanide is yellow), add starch indicator, and complete the titration. While the solution is awaiting titration carry out a blank on the ferrocyanide solution, using the same quantities of reagents, but without necessarily excluding air. Subtract the blank figure (which will probably be appreciable) from the titration, in order to obtain the equivalent of the vanadium present.

1 ml. 0·01 N sodium thiosulphate≡0·5095 mg. V.

Separation of copper from selenium and tellurium. Dissolve 5 g. sample in 80 ml. 10 N sulphuric acid and 30 ml. concentrated nitric acid. Evaporate, fuming strongly to remove traces of nitric acid. Cool, and take up the residue by warming with 100 ml. water. Transfer the solution to a flask, rinsing it in with 100 ml. concentrated hydrochloric acid. Add 1-g. lots of solid sodium hypophosphite till the colour of the solution lightens suddenly, marking complete reduction of the copper, and then add a further 5 g. Fit the flask with a cork carrying a straight 18-inch glass tube, and reflux gently for 15 minutes. Arsenic, selenium, tellurium and a small amount of copper will be precipitated. Filter immediately through a paper-pulp pad, and wash quickly with 100 ml. boiling 6 N hydrochloric acid containing 2 g. sodium hypophosphite. Change the receiver, and boil the filtrate to test for completeness of precipitation.

Wash the precipitate further with four portions of cold 5 per cent ammonium chloride solution, and then dissolve it into a receiver with 15 ml. of a solution containing 5 ml. hydrochloric acid saturated with bromine and 10 ml. water. If necessary pass this solution several times through the filter. Finally wash the filter with 0·5 N hydrochloric acid until the bulk of the filtrate and washings is 100–150 ml. and the filter is white.

Draw air rapidly through the solution to remove bromine, leaving the solution pale lemon-yellow. Add 20 ml. 4 per cent potassium iodide solution, and shake thoroughly. Set aside for 1–1½ hours, shaking occasionally. Filter off the scarlet precipitate on paper pulp. Wash first with 100 ml. 0·5 N hydrochloric acid containing 5 per cent ammonium chloride, add these washings to the filtrate, and reserve the solution for the subsequent determination of tellurium.

Determination of selenium. Wash the selenium with 5 per cent ammonium chloride solution to remove acid, and then five or six times with 5 per cent ammonium nitrate solution to remove chloride, rejecting both lots of wash liquid. Suck the precipitate and filter dry, blow them out into a beaker, and wipe the funnel with a piece

6

of moist filter paper, finally washing it down into the beaker with 10 ml. 10 per cent potassium cyanide solution, and then with hot water. Add enough water to allow the paper pulp to be broken up with a glass rod. Add 0·4 g. sodium nitrate and warm on a steam bath till the precipitate is completely dissolved. Add 25 ml. 8 N nitric acid, stir thoroughly, and boil. Filter into a 600-ml. conical flask, wash the pulp thoroughly with hot water, and add the washings to the filtrate. Evaporate the solution to about 40 ml., cool, add 1 g. urea, and make up to approximately 100 ml. with cold water.

Add 2 ml. 10 per cent aqueous potassium cobalticyanide solution to precipitate traces of copper, shake, and after a lapse of about 1 minute add 10 ml. 4 per cent potassium iodide solution and 5 ml. chloroform, and shake vigorously and frequently over 5 minutes. Titrate the liberated iodine with 0·01 N sodium thiosulphate solution, shaking constantly near the end-point.

If the amount of copper is small or negligible, the final colour after disappearance of the starch-iodide blue is orange, due to precipitated selenium. In the presence of larger amounts of precipitated copper cobalticyanide the selenium appears to be adsorbed by the precipitate.

Standardise the sodium thiosulphate solution against standard iodine to which has been added 20 ml. dilute nitric acid and 1 g. urea.

1 ml. 0·01 N sodium thiosulphate≡0·3948 mg. Se.

Determination of tellurium. Precipitate the tellurium from the filtrate retained for the purpose by passing in sulphur dioxide, with shaking, until the liquid no longer darkens. Boil vigorously for 5 minutes, cool slightly, add a few ml. saturated sulphur dioxide solution, and cool quickly.

When quite cool, filter off the tellurium. Wash the precipitate with cold 5 per cent ammonium chloride solution, and dissolve the tellurium through the filter by bromine-hydrochloric acid as in the selenium determination. Evaporate the filtrate to dryness on a steam bath until no odour of hydrochloric acid remains. Take up the residue in 5 ml. syrupy phosphoric acid and 5 ml. 1 per cent gum arabic solution, shake to disintegrate the filter, and pour into a flask, rinsing the beaker with 50 ml. water in small portions. Add 2 ml. 10 per cent potassium cobalticyanide solution to precipitate copper, followed by 2 g. sodium hypophosphite. Boil vigorously and cool.

When the solution is cold dilute with 150 ml. water, and titrate with 0·01 N iodine solution. Towards the end of the titration when the solution is very pale brown add 10 ml. benzene and shake vigorously. Complete the titration, adding not more than 3–4 drops at a time and shaking thoroughly after each addition, till the benzene layer just turns pink, and the aqueous layer is practically colourless. It is not possible to use starch in this titration owing to the slowness of the iodine-tellurium reaction.

$$\text{1 ml. 0·01 N iodine} \equiv 0·6381 \text{ mg. Te.}$$

[1] B. S. Evans, *Analyst*, 1943, **68**, 67.
[2] B. S. Evans and D. G. Higgs, *ibid.*, 1945, **70**, 158.
[3] B. S. Evans, *ibid.*, 1938, **63**, 870.
[4] *Idem, ibid.*, 1942, **67**, 346.

B. SELECTIVE

1. trans-DICHLORO-bis-ETHYLENEDIAMINECOBALT-III CHLORIDE

Antimony-V can be determined gravimetrically by precipitation from concentrated hydrochloric acid solution with *trans*-dichloro-*bis*-ethylenediaminecobalt-III chloride,

$$trans\text{-}[Co(H_2N.CH_2.CH_2.NH_2)_2Cl_2]Cl,$$

the precipitate being dichloro-*bis*-ethylenediaminecobalt-III hexachlorostibnate.[1] This compound is insoluble both in water and in hydrochloric acid. It is anhydrous, and can be dried without loss or decomposition at 110° C. The precipitate is, however, very soluble in acetone, and somewhat soluble in ethanol and ether, so that washing with an organic solvent, followed by air-drying, cannot be employed.

Precipitation of 1–50 mg. antimony is complete in 50 ml. solution after 2 hours, or in 100 ml. solution after 6 hours. Excess of reagent has no apparent effect.

Either chlorine or nitric acid may be used to convert the antimony to antimony-V for the precipitation. Up to 200 mg. of arsenic, copper or zinc, up to 100 mg. of cadmium, iron, mercury or tin and up to 10 mg. bismuth have no effect. Lead in any amount interferes.

Attempts to complex interfering elements were not successful, but by using a suitable separation procedure antimony may be determined satisfactorily in white metal.

Procedure

Gravimetric determination of antimony. Adjust the volume of the solution in concentrated hydrochloric acid, containing 1–50 mg. to 50 ml. Oxidise the antimony either by bubbling chlorine through the solution or by adding about 3 ml. concentrated nitric acid, warming for a few moments and cooling quickly. Add 0·2 g. *trans*-dichloro-*bis*-ethylenediaminecobalt-III chloride hydrochloride dissolved in 10 ml. 2 N hydrochloric acid, set aside for 2 hours with occasional stirring, and filter on an X4 sintered-glass filter crucible. Wash with water until the washings are colourless, and dry at 100°–110° C. to constant weight (about 1 hour).

$$\text{mg. ppt.} \times 0\cdot2098 \equiv \text{mg. Sb.}$$

Determination of antimony in white metal. Dissolve 1 g. alloy in 40 ml. 8 N nitric acid in a porcelain basin. Add 3 ml. concentrated sulphuric acid, evaporate to fuming, and cool. Wash down the sides of the basin with distilled water, repeat the evaporation, and cool. Add 50 ml. distilled water, stir well, and set aside for 1 hour. Filter off lead sulphate on a sintered-glass filter crucible, wash the precipitate with a little dilute sulphuric acid (1 : 20), and dissolve out the lead sulphate with concentrated ammonium acetate solution poured through the filter. Wash the filter with a mixture of 10 ml. concentrated hydrochloric acid and 5 ml. concentrated nitric acid, and combine this washing and the washings of the lead sulphate precipitate with the original solution.

Add 10 g. citric acid dissolved in 20 ml. water, and make the solution slightly alkaline with ammonia. Treat a saturated solution of potassium cyanide with bromine until it gives no colour with sodium nitroprusside, and add this solution dropwise to the ammoniacal solution until the blue colour is discharged, and then add a further 30 ml. Follow this with 75 ml. 20 per cent ammonium chloride solution, and 10 g. sodium dithionate, and heat just at boiling point on a steam bath for 1 hour. Add a further 5 g. sodium dithionate, cool the solution in running water, and filter through a paper-pulp pad. Wash the precipitated antimony with a cold solution containing 20 ml. saturated potassium cyanide solution, 4 g. ammonium chloride and 2 g. sodium dithionate in 400 ml.

Transfer the filter to a beaker and cover it with 100 ml. concentrated hydrochloric acid containing some chlorine. Break up the pulp and stir until all the antimony has dissolved. Filter off the paper pulp and complete the antimony determination as already described.

Synthesis of the Reagent[2]

Add 600 g. 10 per cent solution of ethylenediamine with stirring to a solution of 160 g. cobalt chloride hexahydrate in 500 ml. water in a 2-litre beaker. Draw air vigorously through the solution for 10–12 hours. Add 350 ml. concentrated hydrochloric acid, and evaporate on the steam bath to a volume of about 750 ml., when a crust forms on the surface of the liquid. Allow the cool solution to stand overnight. Filter off the bright green plates of the hydrochloride. Redissolve these in water, add a large excess of concentrated hydrochloric acid, and evaporate until crystallisation begins. Wash with ethanol and ether, and air-dry.

If dried at 110° C., hydrogen chloride is lost and the solid crumbles to a dull green powder. However, the hydrochloride may be used directly for the preparation of the reagent solution.

[1] R. Belcher and D. Gibbons, *J.C.S.*, 1952, 4775.
[2] J. C. Bailar, *Inorganic Syntheses*, 1946, **2**, 223.

2. DI-*cyclo*-HEXYLTHALLIUM-III SULPHATE

Several di-organo-thallium-III compounds have been examined[1] as precipitants for the nitrate ion, and of those examined the di-*cyclo*-hexyl derivative was found to be the most suitable. This forms a soluble sulphate, perchlorate and acetate, all of which may be used as reagents; and the nitrate is so insoluble that it may be washed without serious loss. It is more readily filtered than nitron nitrate.

The determination may be completed either gravimetrically or titrimetrically. Two titrimetric procedures are possible: in one of these the nitrate solution is titrated with reagent to the flocculation point; in the other method the thallium is precipitated with excess oxalate, the excess being titrated with permanganate. As yet another possibility, the precipitate may be decomposed with fuming nitric acid and the thallium then determined by a standard procedure.

The reagent is stored as the carbonate, and this is dissolved in sulphuric, perchloric or acetic acid before use. Sulphuric acid is normally used, but if there are interferences acetic acid should be substituted.

For gravimetric determination the reagent should be dissolved in sufficient acid to allow 1 mole acid in excess for each mole of salt. The thallium reagent is more soluble under these conditions and the complex nitrate crystallises out much better than from neutral solution. The solution should, however, be neutral for the oxalate-permanganate titrimetric method, or if fluoride is present.

In neutral solution the following ions interfere: bromide, carbonate, chloride, chromate, cyanate, cyanide, ferricyanide, ferrocyanide, iodide, nitrite, oxalate, permanganate, sulphide, sulphite, tartrate, thiocyanate; at higher concentrations permanganate is reduced to manganese dioxide. In strong acid solution (pH 0–1·0), the ions carbonate, chromate, nitrite, permanganate, sulphite and tartrate do not give precipitates, and most of the remaining interfering ions can be removed by precipitation with silver fluoride or silver acetate.

Solutions required

Reagent solution. Dissolve sufficient di-*cyclo*-hexylthallium-III carbonate in dilute sulphuric acid or acetic acid to give a 0·05 M solution.

Procedures

Gravimetric determination of nitrate. Adjust the solution, containing 50–100 mg. nitrate, to a volume of 50–100 ml., and boil. Add 50–100 ml. reagent solution and cool to room temperature. Filter through an X3 sintered-glass filter crucible, but if possible do not suck the precipitate dry. Wash two to five times with 10 ml. ice-cold water. At the end of the washing the filtrate should not be turned cloudy by chloride ions. Dry the precipitate at a temperature not exceeding 150° C. and weigh as $C_{12}H_{22}TlNO_3$. Measure the total volume of filtrate and washings in ml. and divide by 150 (150 ml. water dissolves 1 mg. precipitate). Add the quotient in mg. to the weight of precipitate found.

$$mg. \; ppt. \times 0·1456 = mg. \; HNO_3.$$
$$mg. \; ppt. \times 0·1433 = mg. \; NO_3^-.$$

Titrimetric determination of nitrate. Boil the neutral solution which should be free from cations which give precipitates with oxalate. Add exactly 100 ml. 0·05 M reagent solution. Cool, add 50 ml. 0·1 N oxalic acid solution, and filter both precipitates off together. Wash as in the gravimetric determination. Acidify the filtrate with sulphuric acid and titrate the excess oxalic acid with 0·1 N potassium permanganate solution. The permanganate solution consumed corresponds to the nitrate ion originally present, since the amount of reagent solution added is equivalent to the amount of oxalic acid solution added. Because of the instability of the reagent it is advisable to run a blank. The difference between the titration figure and the blank then corresponds to the nitrate content.

1 ml. 0·1 N potassium permanganate≡6·3016 mg. HNO_3.

1 ml. 0·1 N potassium permanganate≡6·2008 mg. NO_3^-.

Synthesis of the reagent

Treat a 0·1 N ethereal solution of thallium-III. chloride with an approximately 0·3 N Grignard solution of *cyclo*-hexyl chloride and magnesium in ether. To the resulting mixture of thallium-I chloride and di-*cyclo*-hexylthallium-III chloride add excess of silver fluoride or acetate solution acidified with sulphuric acid. Precipitate the excess of silver with hydrogen sulphide. Boil the filtrate to expel hydrogen sulphide, and add sodium carbonate. Filter, and wash the precipitate with water until the washings no longer give a reaction for thallium with iodide. Dissolve the complex in sulphuric acid, reprecipitate as the carbonate, and repeat the treatment. Dry at 105° C.

The purity of the reagent may be checked by boiling 1 g. carbonate with 10 ml. 50 per cent acetic acid until no more carbon dioxide is evolved, and diluting the solution to 100 ml. with water. Only pure white precipitates should be formed with hydrogen sulphide or with potassium iodide. Silver acetate or barium acetate should give no precipitate.

[1] H. Hartmann and G. Bäthge, *Angew. Chem.*, 1953, **65**, 107.

3. HEXA-AMMINOCOBALT-III BROMIDE

Hexa-amminocobalt-III bromide, in an aqueous solution containing the appropriate amount of miscible organic solvent, precipitates sulphate quantitatively as the mixed halide-sulphate $(NH_3)_6CoBrSO_4$, this compound forming as a very fine reddish yellow precipitate. Such precipitation takes place in a solution so dilute that barium sulphate produces no precipitate. The solubility of the mixed halide-sulphate in water at $20°C$. is 0.1055 moles per litre, corresponding to a sulphate content of about 1 g. per litre. This figure is considerably reduced by the added organic solvent.

The final determination can be carried out by drying and weighing the precipitate; by redissolving the precipitate in hot water and determining the bromide argentometrically;[1] or by redissolving the precipitate in dilute acid followed by a colorimetric comparison with standard solutions of $(NH_3)_6CoBrSO_4$ or of $(NH_3)_6CoCl_3$. The titrimetric finish gives rather a poor end-point. Using a gravimetric finish the average error on a sample containing 30.0 mg. SO_4^- is 0.05 mg.

The test solution should be neutral or weakly acidic, with a pH of not less than 2. Alkali chloride, nitrate or perchlorate, or the corresponding ammonium salts, when present in moderate amounts, do not interfere. Sodium chloride or magnesium chloride present in large amount give rise to errors. Alkali nitrate or chloride should therefore not be present in amounts greater than six or seven times, or magnesium nitrate or chloride in amounts greater than four times that of the corresponding sulphate. Calcium, if present, is precipitated, and other multivalent cations interfere. It is therefore recommended that cations be removed by stirring with a suitable base-exchange agent, such as Wolfatit, before the actual precipitation.

Phosphate may interfere since it gives an insoluble phosphate, $(NH_3)_6CoPO_4$, in neutral or weakly acid solutions. If this ion is present the pH should be adjusted so that a yellow colour is formed with bromophenol blue (pH $2.3–2.5$), when the phosphate no longer precipitates.

Solution required

Reagent solution. Dissolve 1.5 g. hexa-amminocobalt-III-bromide and 0.25 g. ammonium bromide in 100 ml. 0.1 N hydrochloric acid by gentle warming. Cool, and add 600 ml. ethanol or *iso*-propanol; or 670 ml. methanol or acetone, and make to 1 litre with water. If the solution goes cloudy on standing, add a little animal charcoal or paper pulp, and filter through paper without suction. This reagent solution is stable.

Procedures

Gravimetric determination of sulphate. Add the appropriate organic solvent to the test solution to bring it to the same proportion by volume as the reagent solution. If the solvent concentration is higher than that in the test solution, reagent may be precipitated when the two solutions are mixed. Warm the test solution quickly to 40° C. to avoid loss of solvent, and add 1 ml. reagent solution for each mg. $SO_4^=$ present. Cool the mixture in running water or in ice, and filter through an X4 sintered-glass filter crucible. The filtrate should be yellow, indicating that an excess of the reagent has been added.

Wash the precipitate with a methanol-water or an acetone-water (2 : 1) wash liquid. Wash finally with pure ethanol, and air-dry with filtered air, or dry at 75°–80° C. in an oven.

The precipitate should not be allowed to overheat or it will decompose. Such decomposition will be indicated by darkening.

$$\text{mg. ppt.} \times 0\cdot2849 \equiv \text{mg. } SO_4^=.$$

Titrimetric determination of sulphate. Proceed as in the gravimetric determination to the stage of washing the precipitate with methanol-water or acetone-water. Redissolve the washed precipitate in hot water, and add 1·5 ml. 5 per cent potassium chromate solution for each 50 ml. solution. Titrate with 0·1 N silver nitrate solution from a burette graduated in 1/100 ml. Observe the end-point by comparison with a standard which has previously been titrated accurately.

$$1 \text{ ml. } 0\cdot1 \text{ N silver nitrate} \equiv 9\cdot606 \text{ mg. } SO_4^=.$$

Synthesis of hexa-amminocobalt-III bromide[2]

Add 24 g. cobalt carbonate slowly to 100 ml. 45 per cent hydrobromic acid, followed by 2 g. activated charcoal and 120 ml. concentrated ammonia. Disregarding any precipitate which may appear, add 40 ml. 30 per cent hydrogen peroxide slowly with stirring. Allow the vigorous effervescence to cease. Heat the mixture for 5 minutes on a steam bath, and then set it aside for half an hour. Filter, discarding the filtrate, and wash the mixture of carbon and product with a little cold water. Treat the mixture with 900–1000 ml. water containing sufficient hydrobromic acid to give a slight acid reaction. Heat on the steam bath until all the product has dissolved, and filter the solution to remove the charcoal. Add 50 ml. 45 per cent hydrobromic acid to the hot solution, and cool slowly to 0°C. Filter, wash the precipitated product with ice-cold water and ethanol, and dry at 100°C.

Yield about 64 g. (80 per cent).

[1] C. Mahr and K. Krauss, *Z. anal. Chem.*, 1948, **128**, 477.
[2] J. Bjerrum and J. P. McReynolds, *Inorganic Syntheses*, 1946, **2**, 219.

4. OCTA-AMMINO-μ-AMINO-μ-NITRODICOBALT-III NITRATE

Of a range of five co-ordination compounds of cobalt examined by Belcher and Gibbons,[1] octa-ammino-μ-amino-μ-nitrodicobalt-III nitrate,

$$\left[(NH_3)_4 . Co \underset{NO_2}{\overset{NH_2}{<\,>}} Co . (NH_3)_4 \right] (NO_3)_4 . H_2O.$$

proved to be the most promising as a reagent for sulphate.

The sulphate is more soluble (22·4 mg. per l.) than barium sulphate (2·3 mg. per l.) so that in order to obtain the best results it is necessary to work in controlled volumes of solution. The solubility is, however, much more favourable than that of benzidine sulphate (98 mg. per l.).

The particular advantages over barium sulphate are that co-precipitation errors appear to be negligible, and that the nitrate ion in particular does not interfere. No interference is found from calcium, potassium, sodium, chloride, fluoride or peroxide.

Interference from iron-III and aluminium in neutral solution owing to hydrolysis can be overcome to some extent by increasing the acidity. A preferable treatment, however, is to add sufficient ethylenediaminetetra-acetic acid to complex these ions. Phosphate interferes at a pH greater than pH 5·0, but this can be overcome by adjusting the pH to 4·0–5·0 if phosphate is present. In this range of acidity the sensitivity of the reagent is unaffected.

The precipitate may be dried at 110°C. to the anhydrous compound, or it may be washed with organic water-miscible solvents and air-dried as the dihydrate. Acetone is found to give better recoveries than methanol or ethanol.

Solutions required

Ethylenediaminetetra-acetic acid. Di-sodium salt, 0·5 M aqueous.
Reagent solution. 1 per cent aqueous. This is stable for at least a week.

Procedures

Gravimetric determination of 1–15 mg. sulphate. Evaporate the solution to a bulk that will give rise to a final volume of 20 ml. when all reagents have been added. Add 2 ml. ethylenediaminetetra-acetic acid solution for each mg.-atom of aluminium or iron-III present. Adjust the solution to neutrality. If phosphate is present, add 1 ml. 2 M hydrochloric acid. Add sufficient reagent solution to produce a two- or three-fold excess (0·3 ml. per mg. $SO_4^=$), and sufficient acetone to give a concentration of this solvent of 25 per cent (v/v). Stir occasionally over a period of 4 hours.

Filter off the precipitate on an X4 sintered-glass filter crucible, and wash first with 25 per cent acetone and then with pure acetone. Dry in a current of filtered air, place in the balance-case for 30 minutes, and weigh as the dihydrate.

Alternatively dry in an oven at 110° C. and weigh as the anhydrous compound.

Oven-dried: mg. ppt. $\times 0·378 \equiv$ mg. $SO_4^=$.
Air-dried: mg. ppt. $\times 0·353 \equiv$ mg. $SO_4^=$.

Gravimetric determination of 15–20 mg. sulphate. Proceed as described above, but using a final volume of 50 ml.

Gravimetric determination of 50–100 mg. sulphate. Proceed as for 15–20 mg. sulphate, but filter off the precipitate after 30 minutes.

[1] R. Belcher and D. Gibbons, *J.C.S.*, 1952, 4216.

CHAPTER IV

ORGANIC REAGENTS

1. ALIZARIN BLUE

A LIZARIN BLUE

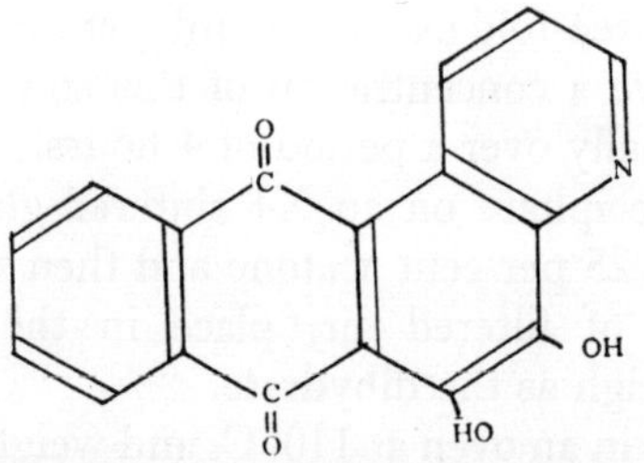

is known as a specific reagent for copper-II ion. In strongly acid solution only this ion forms a precipitate, which is blue and crystalline, with a pyridine solution of the reagent, although in ammoniacal solution cadmium, nickel and zinc yield precipitates which dissolve in acids. The copper complex is not decomposed by sulphide or cyanide ions as is usual with most other water-insoluble copper complexes.[1]

The main obstacle to the development of a gravimetric method for copper using this reagent lies in the fact that the reagent is only soluble to the extent of 0·2 per cent in pyridine. Accordingly when the pyridine solution is added to an acid solution containing copper-II, the reagent is also precipitated. However, when the contaminated copper precipitate is treated first with acetic anhydride and then with pyridine, the reagent is completely removed, whereas the copper complex remains unaltered. No occlusion of the reagent occurs. The removal of the dye is thought to be possible because of acetylation of the phenolic groups, since the resulting solution does not react with copper-II, but when it is evaporated and treated with pyridine the reactivity is re-established.

76

Because of the voluminous nature of the precipitate the amount of copper present should not exceed 2 mg.

When the reagent is added to a solution of copper-I cyanide the copper-II compound is formed. The system

$$Cu_2(CN)_2 \rightleftharpoons Cu_2{}^{++} + 2CN^- \qquad (1)$$

undergoes autoxidation according to the reaction

$$Cu_2{}^{++} + (O) + 2H^+ \longrightarrow 2Cu^{++} + H_2O. \qquad (2)$$

The copper-II ions formed in (2) react with the reagent, disturbing the equilibrium (1), which is immediately re-established. In this way all the copper-I cyanide is transformed into the complex copper-II compound. Copper-I cyanide is produced from solutions containing cuprocyanide ions by adding formaldehyde. Copper may therefore be determined in solutions containing cyanide.

Solution required

Reagent solution. A saturated solution of alizarin blue in pyridine. Dissolve an excess of reagent in hot pyridine, and filter after cooling to room temperature, to obtain a stable solution.

Procedures

Gravimetric determination of copper. Add 20 ml. 6 N sulphuric acid to 20 ml. aqueous solution containing not more than 2 mg. copper-II. Heat the solution, and add the reagent solution slowly. The reaction is slow at first, but once initiated it continues more rapidly. When precipitation is complete the supernatant solution should be reddish-blue and should be distinctly acid. Set the vessel aside for 20–60 minutes, and filter on a fine sintered-glass filter crucible, using 3 N sulphuric acid to effect the transference.

Place the crucible in a small beaker filled to half the height of the crucible with acetic anhydride. Cover the beaker with a clock glass and heat to 60°–80° C. for 30 minutes. Filter the acetic anhydride from the crucible, and wash with hot pyridine until the filtrate is no longer blue. Wash the precipitate with ethanol, dry at 100° C. for 30 minutes, and weigh.

$$\text{mg. ppt.} \times 0.0991 = \text{mg. Cu.}$$

Determination of copper in presence of cyanide. To the solution containing not more than 2 mg. copper as the complex cuprocyanide ion in a volume of 20 ml. add a few drops formaldehyde. Acidify with 20 ml. 6 N sulphuric acid and heat to boiling to expel excess formaldehyde. Continue the determination as described above.

[1] F. Feigl and A. Caldas, *Anal. Chim. Acta*, 1953, **8**, 339.

2. 5:6-BENZOQUINOLINE

Trioxalatogermanic acid is too unstable to be isolated, but it forms insoluble salts with quinine and strychnine, and with 5 : 6-benzoquinoline (β-naphthoquinoline),[1]

The complex with the last mentioned reagent has been proposed[2] as providing a gravimetric method for the determination of germanium.

The 5 : 6-benzoquinoline trioxalatogermanate is easy to filter and wash, so that contamination is slight. However, some reagent is co-precipitated, and in addition the precipitate loses weight slowly, only coming to constant weight after 30 hours in a desiccator. Consequently the high factor cannot be utilised, and the precipitate must be ignited to the oxide for final weighing.

Tin, titanium and zirconium also give complex oxalates under similar conditions. That with tin resembles the germanate, while the titanium and zirconium precipitates appear to be more insoluble and flocculent. Other metals giving insoluble oxalates interfere. High concentrations of sodium chloride prevent complete precipitation of the complex, and may inhibit precipitation altogether.

Reagent solution

Dissolve 10 g. 5 : 6-benzoquinoline and 5 g. oxalic acid in 50 ml. water. Heat to aid solution, filter while hot and dilute to 500 ml. with water.

Procedure

Determination of germanium. Dilute a measured amount of germanium solution to 400 ml., add 4 g. oxalic acid, and heat the solution to favour the formation of trioxalatogermanic acid. To the hot solution add 25 ml. reagent solution, and allow to cool to room temperature. The germanate precipitates in long needles. Set aside overnight, filter, and wash with a dilute solution of 5 : 6-benzoquinoline-oxalate reagent. Ignite in a platinum crucible at 800° C. The residue, germanium oxide, GeO_2, should be pure white.

$$mg. \ ppt \times 0 \cdot 6940 \equiv mg. \ Ge.$$

[1] A. Tchakirian, *Ann. Chim.*, 1939, **12**, 415.
[2] H. H. Willard and C. W. Zuehlke, *Ind. Eng. Chem. Anal.*, 1944, **16**, 322.

3. BENZO-1 : 2 : 3-TRIAZOLE

Benzo-1 : 2 : 3-triazole,

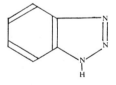

a white crystalline solid, m.p. 96°–97° C., which is soluble in water, ethanol, benzene and chloroform, forms a precipitate with copper which has the formula $(C_6H_4N_3)_2Cu$,* and which affords a method for the gravimetric or titrimetric determination of copper.[1]

The precipitation must be carried out in a narrow pH range, 7·0–8·5, in a solution containing tartaric and acetic acids. The reagent is employed as a 2 per cent aqueous solution, this being almost saturated at room temperature.

* However, cf. C. Duval, *Inorganic Thermogravimetric Analysis* (Amsterdam, 1952), p. 264, where the composition is stated to be variable.

Where elements which also precipitate are absent, the complex may be dried and weighed directly. Preferably, and always where interfering elements are present, as is the case with cast irons and steels, the complex is ignited to the oxide, dissolved in nitric acid, and the copper is then determined iodometrically.

Cadmium, cobalt, iron-II, nickel, silver and zinc give similar precipitates, and so interfere with the gravimetric determination. If large amounts of these elements are present a bulky precipitate, which is difficult to filter and ignite, may render even the titrimetric determination impracticable.

Aluminium, antimony, arsenic, chromium, iron-III, molybdenum, selenium and tellurium do not interfere, and so a method is afforded for the determination of copper in the presence of a number of elements which normally interfere in the iodometric determination as carried out after a sulphide precipitation.

Procedures

Gravimetric determination of copper. To the solution containing copper add 7 g. tartaric acid and 10 ml. glacial acetic acid. Adjust the volume to 200 ml., and make just alkaline to litmus with ammonium hydroxide solution (sp. gr. 0·90). Add 50 ml. 2 per cent aqueous reagent, and digest on a hot plate until the precipitate coagulates. Filter through a sintered-glass filter crucible, dry for 2–3 hours at 135°–140° C., cool in a desiccator, and weigh as $(C_6H_4N_3)_2Cu$.

$$mg. \; ppt. \times 0·2122 \equiv mg. \; Cu.$$

Titrimetric determination of copper in cast irons and steels. Place the weighed sample (1–10 g.) in a 400-ml. beaker, add 25 ml. nitric acid (sp. gr. 1·13) per gram of sample, and heat to dissolve, finally expelling oxides of nitrogen. If more than 0·5 per cent chromium is present, dissolve in hydrochloric acid and subsequently oxidise the iron by nitric acid. Add 7 grams tartaric acid per gram of iron, and 10 ml. glacial acetic acid. Bring the volume to 200 ml., adjust the acidity, and carry out the precipitation as already described.

When the precipitate has coagulated, add a small plug of paper pulp, stir, and filter through a Whatman No. 40 filter paper. Wash the precipitate thoroughly with cold water, transfer to a crucible, and ignite in a muffle at red heat. Transfer the copper oxide to a 250-ml. beaker, add 10 ml. nitric acid (sp. gr. 1·42), and evaporate just to dryness. Add 20 ml. water, and then add dropwise sufficient 6 per cent sodium hydroxide solution to produce a permanent precipitate. Add 10 ml. glacial acetic acid and boil till the precipitate dissolves. Cool to room temperature, add 6 ml. 50 per cent potassium iodide solution, and titrate the liberated iodine with standard sodium thiosulphate solution, using starch as indicator.

1 ml. 0·1 N sodium thiosulphate≡6·357 mg. Cu.

[1] J. A. Curtis, *Ind. Eng. Chem. Anal.*, 1941, **13**, 349.

4. N-BENZOYLPHENYLHYDROXYLAMINE

N-Benzoylphenylhydroxylamine

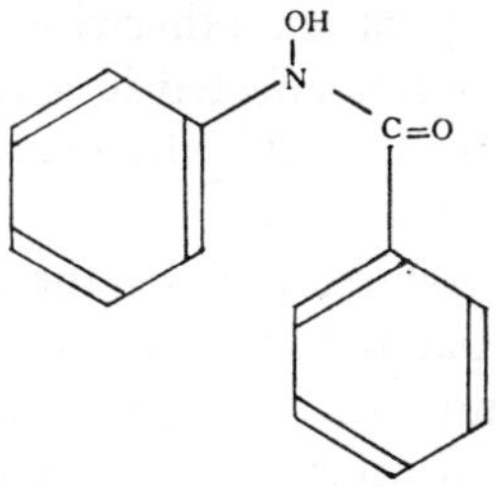

may be used for the gravimetric determination of copper, iron, aluminium and titanium.[1] It has some advantages over cupferron, the ammonium salt of N-nitrosophenylhydroxylamine:

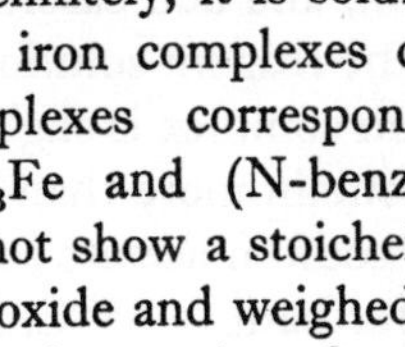

It can be preserved indefinitely, it is soluble in hot water, and the copper, aluminium and iron complexes can be weighed directly, since they give complexes corresponding to the formulae (N-benz)$_2$Cu, (N-benz)$_3$Fe and (N-benz)$_3$Al respectively. The titanium complex does not show a stoicheiometric composition, but may be ignited to the dioxide and weighed as such.

Arsenate and arsenite do not interfere in the determination of copper if precipitation of copper phosphate and other insolubles is prevented by the addition of a small amount of Rochelle salt. Iron

7

and aluminium, however, cannot be separated from phosphate. Molybdate, tungstate and vanadate ions give precipitates with the reagent.

Beryllium, cadmium, cobalt, lead, manganese, mercury-II, nickel, uranium-VI and zinc do not give precipitates with the reagent at pH 4·0. Tin, titanium and zirconium are precipitated. Chromium-III interferes with the precipitation of iron. Aluminium, copper and iron cannot be determined in the presence of one another.

The effective pH ranges for the precipitation of these elements are copper, pH 3·6–6·0 : iron, pH 3·0–5·5 : and aluminium, pH 3·6–6·4. Below these ranges precipitation is incomplete, and above, slightly high results are obtained.

In the determination of titanium the solution must be kept below 25°C. during the precipitation, otherwise a gummy substance is formed. The precipitation is effected in a volume of about 400 ml. containing 5–20 ml. concentrated hydrochloric acid. Low results are obtained when more acid is present.

Aluminium does not interfere under these conditions. Titanium cannot, however, be separated from iron or from phosphate. Apart from noting that tin and zirconium precipitate in acid solution, no indication is given as to other elements which interfere. It may, perhaps, be inferred that they are the same as those which interfere with the copper-iron-aluminium precipitation, although conditions are somewhat different.

Procedures

Gravimetric determination of copper, iron or aluminium. The solution should contain about 0·025 g. copper, 0·015 g. iron or 0·01 g. aluminium. Add 5 ml. 1·08 N sulphuric acid, dilute to 400 ml., and heat to boiling. Dissolve 1·75 times the theoretical amount of reagent in 15–20 ml. ethanol, warm, and add carefully to the hot solution of the metal so that none of the reagent solution falls on the sides of the vessel. Since the aluminium complex is slightly soluble in ethanol the final volume should not contain more than 5 per cent ethanol.

Add 10 ml. 10 per cent sodium acetate solution to raise the pH to about 4. Heat on the water bath to 65°C. (1 hour for copper and 2 hours for iron and aluminium precipitates) with occasional stirring. Do not allow the temperature to rise above 70°C. Filter the precipitate on an X4 sintered-glass filter crucible, wash thoroughly with hot water (water at 45°C. for aluminium) and dry at 110°C. to constant weight.

$$\text{mg. ppt.} \times 0.1303 \equiv \text{mg. Cu.}$$
$$\text{mg. ppt.} \times 0.08036 \equiv \text{mg. Fe.}$$
$$\text{mg. ppt.} \times 0.04064 \equiv \text{mg. Al.}$$

Gravimetric determination of titanium. Dilute the solution containing about 0·1 g. titanium to 400 ml. Neutralise with ammonia solution and add 5 ml. concentrated hydrochloric acid. Precipitate the titanium complex by slow addition of a 10 per cent ethanolic solution of the reagent (about twice the theoretical amount) while stirring. Because of the slight solubility of the complex in ethanol the final solution should not contain more than 5 per cent ethanol.

Allow the precipitate to stand for 45 minutes with occasional stirring. Prepare a wash solution containing 3 ml. concentrated hydrochloric acid per litre 0·1 per cent aqueous reagent. Filter off the complex on paper and wash with this solution. Ignite carefully in a platinum crucible to constant weight and weigh as TiO_2.

$$\text{mg. ppt.} \times 0.5995 \equiv \text{mg. Ti.}$$

Synthesis of N-Benzoylphenylhydroxylamine

Dissolve 30 g. phenylhydroxylamine in 1200 ml. warm water, and filter. Cool the filtrate, add a little sodium bicarbonate, and then add 45 g. benzoyl chloride dropwise whilst stirring vigorously. Add about 30 g. sodium bicarbonate in small amounts during the course of this procedure, to keep the mixture fairly alkaline. Stir for 90 minutes, filter, and wash the solid with water. This is a mixture of the mono- and di-benzoyl derivatives, contaminated with benzoyl chloride.

Triturate the solid with 10 per cent sodium bicarbonate solution in a porcelain mortar for 30 minutes, filter, and wash with water. This treatment removes the benzoyl chloride. Treat the solid with aqueous ammonia (sp. gr. 0·88) to dissolve the mono-benzoyl derivative. Filter, and add the filtrate to a slight excess of dilute sulphuric acid cooled with an ice-salt mixture. Filter the product which separates out, and purify it by recrystallisation from ethanol.

The product is very slightly soluble in cold water, but dissolves to an extent of 0·5 per cent in hot water. Its melting point is 121°–122°C. It is stable in heat, light and air.

[1] S. C. Shome, *Analyst*, 1950, **75**, 27.

5. 4-CHLORO-4'-AMINODIPHENYL

4-Chloro-4'-aminodiphenyl

gives a sulphate which is appreciably more insoluble than that of diaminotolane (p. 88), permitting amounts of sulphate down to 1 mg. to be determined satisfactorily. The reagent is used exactly like benzidine. Apart from the advantage in accuracy, nitrate does not interfere.[1]

The reagent is best used at pH 1·0–2·0. Above this range the precipitate is finely divided and difficult to filter, and below it the hydrochloride tends to precipitate out, and difficulty is experienced in washing out the excess of reagent. Acetate, bromide, chloride, citrate, iodide, nitrate or tartrate do not interfere in a twenty-fold excess. Oxidising agents (except hydrogen peroxide) decompose the reagent. Copper, magnesium and zinc form soluble complexes, but do not interfere up to a ten-fold excess. Alkali metals, calcium, chromium-III or iron-III do not interfere up to a twenty-fold excess. Aluminium forms an insoluble complex, but the interference is easily eliminated by masking with tartaric acid. Oxalate, phosphate, selenate and tellurite interfere by forming insoluble salts. Phosphate could possibly be removed as calcium or magnesium phosphate.

Solutions required

0·24 per cent reagent solution. Dissolve 2·4 g. 4-chloro-4'-aminodiphenyl hydrochloride, with warming, in about 800 ml. water containing 50 ml. N hydrochloric acid. Filter hot and dilute to 1 litre.

0·48 per cent reagent solution. Treat 4·8 g. of the hydrochloride as already described. On standing, this solution deposits some hydrochloride, so that warm freshly prepared reagent should be used.

Sodium hydroxide solution. 0·1 N and 0·05 N.

Mixed indicator solution. Mix three parts 0·04 per cent aqueous phenol red (sodium salt) and two parts 0·04 per cent aqueous bromothymol blue (sodium salt).

Procedures

Titrimetric determination of 2·5–25 mg. sulphate. Adjust the test solution to a volume of 1–10 ml. and to pH 7·0 in a conical beaker, and add a small amount of paper-pulp slurry. Add 5 mg. tartaric acid for each mg. aluminium present. Add sufficient 0·24 per cent reagent solution to give a two-fold excess (6 ml. per mg. $SO_4^=$). Set aside for 15–20 minutes. Filter on a paper-pulp pad on a Witt plate fitted in a conical glass filter funnel and return 5–10 ml. mother liquor to the precipitation vessel for use in transferring any appreciable amounts of precipitate which may have been retained on the walls. Wash the vessel well with two or three 5-ml. portions of distilled water, allowing the filter to drain before each addition of wash liquid. Then wash with water saturated with precipitate until the washings are chloride-free. Drain under gentle suction.

Transfer the precipitate and the upper layer of paper pulp to the precipitation vessel, using a stainless-steel spatula. Use the lower layer of paper pulp to transfer particles of precipitate adhering to the walls of the funnel. Stir the pulp slurry vigorously to break up any appreciable aggregates of precipitate, since sodium hydroxide reacts very slowly with the precipitate unless it is finely divided. Wash the filter funnel with distilled water, collecting the washings in the pre-cipitation vessel, adjust the volume to 40–50 ml. with water, boil for 1 minute, add 3–4 drops indicator solution, and titrate with 0·05 N sodium hydroxide solution to the first purple colour of the solution. Reheat the solution and continue the titration until the first per-manent purple tinge is obtained. The pulp retains a small amount of precipitate so that the first end-point is false, but provides an excellent guide to the nearness of the true end-point.

$$1 \text{ ml. } 0\cdot05 \text{ N sodium hydroxide} \equiv 2\cdot407 \text{ mg. } SO_4^=.$$

Titrimetric determination of 25–100 mg. sulphate. Adjust the volume of solution to 5–25 ml. and the pH to 7·0. Add 3 ml. warm (40° C.) 0·48 per cent reagent solution per mg. $SO_4^=$ and set aside for 30 minutes. Complete the determination as before, but carry out the final titration with 0·1 N sodium hydroxide solution.

Recovery of reagent

Collect the filtration and titration liquors in a 5-litre flask. When almost full, render the contents alkaline and filter on pulp on a Büchner funnel. Wash the pulp residue well with water, drain, and transfer to a 600-ml. beaker. Stir the pulp with 150 ml. warm methanol, filter, and wash with 200–300 ml. ether in several portions. Distil off the ether and about 50 ml. methanol, and retain the distillate for future recoveries.

When sufficient material has accumulated, treat the methanolic solution of 4-chloro-4'-aminodiphenyl with 50 ml. hydrochloric acid (1 : 1) and dissolve the precipitated hydrochloride by boiling, adding more methanol if necessary. Allow to cool. Filter off the colourless, shimmering leaves of hydrochloride which deposit, and dry. Obtain a further yield by concentrating the filtrate. The recovery is practically quantitative.

[1] R. Belcher, A. J. Nutten and W. I. Stephen, *J.C.S.*, 1953, 1334.

6. *m*-CRESOXYACETIC ACID

A reagent which allows the separation of zirconium from a considerable number of other elements is *m*-cresoxyacetic acid,[1] m.p. 102°–103°C.,

Precipitation of the zirconium is quantitative, and the voluminous precipitate which forms in solutions which are 0·20–0·25 N in hydrochloric acid is easy to handle. Its composition varies somewhat, but corresponds approximately to a basic salt with the formula

$$ZrO(OH).C_6H_4(CH_3)O.CH_2.COO.$$

Because of this variable composition the precipitate is not weighed directly, but is ignited to zirconium oxide.

Aluminium, barium, beryllium, calcium, lanthanons, nickel and the uranyl ion do not interfere. Thorium does not interfere if the acid concentration is above 0·1 N. Chromium, tin-II, titanium and vanadium-IV, which are not themselves precipitated by the reagent in the conditions mentioned are, however, slightly co-precipitated in the presence of zirconium. Co-precipitation of the chromium and vanadium-IV is sufficient to colour the zirconium oxide residue, but the amount is not significant, and with a double precipitation all interference from the four elements is avoided. Iron-III is partially precipitated and sulphate interferes with the zirconium determination.

Procedure

Gravimetric determination of zirconium. Measure out a volume of solution containing not more than 0·1 g. zirconium reckoned as oxide. Add 2 N hydrochloric acid and water to give a volume of 100 ml. and a free acid concentration of 0·4 N. Boil the solution and add 10 g. solid ammonium nitrate. Add 100 g. boiling 2 per cent aqueous reagent solution with constant stirring. Continue to boil for 5 minutes and set aside to cool somewhat. Filter through a Whatman No. 42 filter paper. If chromium, tin-II, titanium or vanadium-IV is present, transfer the precipitate and filter to the original beaker, redissolve by digesting with 5 N hydrochloric acid on a water-bath, adjust the acidity, and reprecipitate and filter as before.

Wash the precipitate with hot 0·1 per cent reagent in 0·2 N hydrochloric acid, and then with water. Ignite, and weigh as ZrO_2.

$$mg. \; ppt. \times 0.7403 \equiv mg. \; Zr.$$

Synthesis of the reagent

The method described is derived from a general method for the preparation of aryloxyacetic acids.[2]

To a mixture of 1·0 g. *m*-cresol with 3·5 ml. 33 per cent sodium hydroxide solution add 2·5 ml. 50 per cent chloracetic acid solution and, if necessary, a little water to dissolve the sodium salt of the cresol. Stopper the vessel loosely and heat for 1 hour on a gently boiling water bath. Cool, dilute, and acidify to Congo red with mineral acid. Extract once with ether. Wash the ether extract once with a little water and remove the product by washing with dilute sodium carbonate solution. Acidify to produce the free acid and recrystallise this from water. Yield 1·01 g.

[1] M. Venkataramaniah and B. S. V. R. Rao, *Analyt. Chem.*, 1951, **23**, 539.
[2] C. F. Koelsch, *J. Amer. Chem. Soc.*, 1931, **53**, 304.

7. 4 : 4'-DIAMINOTOLANE

The sulphate of 4 : 4'-diaminotolane,

$$H_2N-\langle\text{benzene ring}\rangle-C\equiv C-\langle\text{benzene ring}\rangle-NH_2$$

is notably less soluble[1] (59 mg. per l.) than that of benzidine (98 mg. per l.), and therefore 4 : 4'-diaminotolane can be used in place of benzidine in the determination of sulphate.[2] Results are more accurate, and the interference of other ions is less marked. A four and a half times excess of reagent is recommended as being the most suitable amount to use, and the precipitation is best carried out at pH 3·0–4·0. The method is satisfactory for amounts of sulphate greater than 10 mg., and has been tested successfully up to 25 mg. sulphate.

Chromate and phosphate interfere, but chromium-VI may be reduced to chromium-III in which form it does not interfere. Aluminium interferes, but the effect can be overcome by addition of tartaric acid. Iron-III appears to react, but results in its presence are only slightly low. Copper reacts, but the determination of sulphate is not affected. Calcium, magnesium and zinc, and nitrate and perchloric ions up to 100 mg. do not interfere.

A determination may be completed in 45 minutes.

Solution required

Reagent solution. Dissolve 1 g. 4 : 4'-diaminotolane in 5 ml. acetone. Add concentrated hydrochloric acid dropwise till precipitation of the hydrochloride is complete. Filter at the pump and dissolve the solid in 375 ml. water. Filter the solution, and add a few drops acetone or N hydrochloric acid to clear the solution.

This solution does not deteriorate after standing 24 hours, but it is advisable to prepare it afresh for each series of determinations. The reagent solution should not be heated.

Procedure

Titrimetric determination of sulphate. Add to the sulphate solution, in a 100-ml. beaker, 90 ml. reagent for each 25 mg. sulphate, and a little paper pulp. Make the volume up to 100 ml. and adjust to pH 3·0–4·0 by dropwise addition of N hydrochloric acid. After 30 minutes filter through a paper-pulp pad, transferring the precipitate with water. Wash the precipitate with 12–15 ml. water until the washings are chloride-free. Transfer the pad to a 100-ml. beaker, adjust the volume to 30–40 ml. with water, add 5 drops 0·1 per cent phenolphthalein indicator solution, and titrate the boiling solution with standard sodium hydroxide solution to the first appearance of a pink colour.

$$1 \text{ ml. } 0{\cdot}1 \text{ N sodium hydroxide} \equiv 4{\cdot}803 \text{ mg. } SO_4^{=}.$$

Synthesis of the reagent

Suspend 10 g. 4 : 4′-dinitrostilbene in dioxan. Add excess bromine to the warm suspension, and boil until as much as possible has gone into solution. Filter hot, pour the filtrate into an excess of water, and filter off the crude 4 : 4′-dinitrostilbene dibromide.

Suspend 77 g. of this product in 500 ml. methanol, and add 58 g. potassium hydroxide dissolved in 150 ml. methanol to the boiling mixture in the course of 1 hour. Boil for a further 45 minutes, and filter hot. Wash the 4 : 4′-dinitrotolane with ethanol and then with water.

Suspend the crude 4 : 4′-dinitrotolane in 350 ml. ethanol, and add 70 g. zinc dust slowly to the stirred mixture. Cool, and add 220 ml. hydrochloric acid (sp. gr. 1·16) dropwise, keeping the temperature below 20° C. Stir for a further 8 hours at 15°–20° C. Decant from the zinc dust, filter, and wash the solid residue with 50 per cent ethanol. Combine the filtrate and washings, and treat with 100 ml. 25 per cent sulphuric acid. Filter off the precipitate of 4 : 4′-diaminotolane sulphate, and heat with an excess of sodium hydroxide solution. Filter to obtain the 4 : 4′-diaminotolane. Yield 12 g.

The amine may be recrystallised from acetone, but the yield is poor. The m.p. of the recrystallised product is 235° C.

[1] R. Belcher, M. Kapel and A. J. Nutten, *Anal. Chim. Acta*, 1953, **8**, 122.
[2] *Idem, ibid.*, 146.

8. *anti*-1 : 5-DI-(*p*-METHOXYPHENYL)-1-HYDROXYLAMINO-3-OXIMINO-PENT-4-ENE

Tungsten may be precipitated[1] by an ethanolic solution of *anti*-1 : 5-di-(*p*-methoxyphenyl)-1-hydroxylamino-3-oximino-pent-4-ene, m.p. 156°–157° C.

This compound crystallises from ethanol as small yellow needles and gives yellow solutions in ethanol, acetone, ethyl ether, ethyl acetate, dioxan, acetic acid and benzene. It is insoluble in water and in petroleum ether.

The nature of the complex formed with tungsten is not certain, but one molecule of reagent is used for each molecule of tungstate. The *syn*- form of the compound, which forms colourless needles melting at 217° C., does not react.

The complex is finally ignited to tungsten trioxide and weighed in this form.

The pH of the solution should be less than 1 (about 0·2 N in hydrochloric acid). Sulphuric or nitric acid may be used, but nitric acid stronger than N attacks the reagent. The precipitation should be carried out at room temperature. About twice the theoretical amount of reagent required for precipitation ensures rapid and complete precipitation.

A wide range of cations and anions shows no interference. Brown precipitates are obtained with cerium-IV, gold-III, indium-III and osmate. Tin-II and uranyl ion give orange precipitates and tin-IV gives a yellow precipitate. Copper does not precipitate as long as the solution is kept acid. Molybdenum co-precipitates to a small extent, but no more than when more conventional reagents such as cinchonine are used. Iron-III forms a small amount of brown precipitate, but not enough to interfere seriously even in the analysis of steels containing tungsten.

Solutions required

Reagent solution. Dissolve 0·7 g. reagent in 100 ml. ethanol. Store in a dark glass bottle to prevent darkening of the solution.

Wash solution. Dilute 20 ml. concentrated hydrochloric acid with water to 1 litre and add 1 ml. saturated ethanolic reagent.

Procedure

Determination of tungsten. Adjust the acidity of the solution to about 0·2 N in hydrochloric acid with a total volume of 250 ml. Add 25 ml. reagent solution to the cold tungsten solution slowly and with constant stirring. A deep orange-yellow precipitate forms. Allow this to settle for 2 hours. Test the supernatant liquid with a few drops of reagent. If an orange precipitate forms more reagent is required, but if any precipitate which forms is white or pale yellow this is precipitated reagent, and precipitation is considered to be complete. Filter by decantation through an ashless filter paper, and wash several times with wash solution.

Ignite the paper and residue in a platinum crucible below red heat until all organic matter has been destroyed. Add a few drops of nitric acid, evaporate to dryness, and ignite at 750° C. (not higher) to constant weight. Weigh as WO_3.

The oxide may then be examined for contaminants (silica, molybdenum, iron) in the usual way, and its weight corrected for these if they are found to be present.

In determining tungsten in ores a double precipitation is recommended, taking precautions to recover all traces of tungsten from the filter used after the first precipitation.

$$\text{mg. ppt.} \times 0\cdot7930 \equiv \text{mg. W.}$$

[1] J. H. Yoe and A. L. Jones, *Ind. Eng. Chem. Anal.*, 1944, **16**, 45.

9. DITHIAN

Dithian, m.p. 110°–111° C.,

is the sulphur analogue of dioxan. It is soluble in 95 per cent ethanol. Like dioxan, it forms addition compounds with metallic salts, and the compound with mercuric chloride,

$$HgCl_2 . C_4H_8S_2,$$

has a low solubility in water, precipitating in colourless plates from dilute solutions, and forming a white precipitate from more concentrated solutions of mercuric chloride.

In certain circumstances dithian can be used as a specific reagent for the determination of mercury-II in the presence of other common ions.[1] Mercury-I may also be determined by first oxidising to the mercury-II form. While the method is not highly accurate, being liable to errors of about 0·2–0·3 per cent, the specific action of the reagent offers an advantage.

For the determination the mercury must be present as mercuric chloride, the optimum amount being 50–100 mg. mercury. The solution should be 0·1–0·2 N with respect to hydrochloric acid. Care must be taken to avoid adding excess of the reagent solution, partly because the addition compound is moderately soluble in ethanol, and partly because of the relative insolubility of the reagent itself in dilute aqueous ethanol, leading to precipitation.

Silver, mercury-I and much lead will be absent because of the presence of hydrochloric acid. With the exception of copper, the other common cations, and, in addition, cerium, lithium, titanium and uranium, do not interfere.

Large concentrations of copper give rise to a little co-precipitation of the copper with the mercury addition compound imparting a pink colour to the precipitate. The interference is lessened by reducing the concentration of the hydrochloric acid. Platinum and palladium give precipitates under the same conditions as mercury. Sulphates interfere, since mercuric sulphate also forms an insoluble addition compound.

If precipitation commences immediately on addition of the reagent, it is rapidly completed, but a precipitate which is slow in appearing should not be filtered for 24 hours. The precipitate may be washed with water, and a precipitate which adheres to the wall of the vessel can be loosened by the addition of a drop of detergent solution before filtering.

No serious loss is found on drying the precipitate for 2 hours at 100°C.

Procedure

Determination of mercury. If the solution is acid make it alkaline with ammonia and then neutralise with hydrochloric acid. Filter off any chloride precipitate, and to the filtrate add 2 ml. concentrated hydrochloric acid and adjust the volume with water to about 85 ml. Add sufficient 2 per cent dithian in 95 per cent ethanol to bring the volume to 100 ml. This, under average conditions, provides about a 30 per cent excess of reagent.

Allow the precipitate to settle, preferably for 24 hours, and filter through a tared porous-porcelain crucible. Wash with water, dry for 2 hours at $100°C.,$* and weigh as $HgCl_2.C_4H_8S_2$.

$$mg. ppt. \times 0\cdot5121 \equiv mg. Hg.$$

Synthesis of the reagent[2]

Pulverise 60 g. of sodium hydroxide finely and add to 150 ml. ethanol in a 2-litre round-bottomed flask. Pass in a plentiful supply of sulphuretted hydrogen until the sodium sulphide first formed is converted to sodium bisulphide, and redissolves.

Prepare ten mixtures of 12 ml. ethylene bromide and 45 ml. ethanol, and keep them thoroughly chilled in ice. To each in turn add 6 g. finely ground sodium hydroxide and immediately add the mixture to the sodium bisulphide solution, shaking constantly. When all ten portions of mixture, representing a total of 120 ml. ethylene bromide, 450 ml. ethanol and 60 g. sodium hydroxide, have been added, continue to pass sulphuretted hydrogen for 3 hours through the cooled mixture. Add a few grams ground sodium hydroxide, and steam distil. Yield 24–30 g.

[1] J. B. Schroyer and R. M. Jackman, *J. Chem. Educ.*, 1947, **24**, 146.
[2] J. W. Bouknight and G. M. Smith, *J. Amer. Chem. Soc.*, 1939, **61**, 28.

10. ETHYLENEDIAMINETETRA-ACETIC ACID

Since 1945 increasing attention has been paid to the analytical uses of a series of compounds known as "complexones" (or by a range of other trivial names).[1] The best known, and probably the most useful of these are "complexone I"—nitrilotriacetic acid or ammonia triacetate, $N(CH_2COOH)_3$, and "complexone II" or EDTA—ethylenediaminetetra-acetic acid,

$$(HOOC.H_2C)_2N.CH_2.CH_2.N(CH_2.COOH)_2.$$

Because of the reactive groups which these compounds contain, their outstanding characteristic is an ability to form stable complexes with

* C. Duval, *loc. cit.*, p. 445, states that the precipitate begins to decompose at 97°C.

a wide range of cations. The complex of ethylenediaminetetra-acetic acid with calcium may be instanced as a suitable example of these complexes:

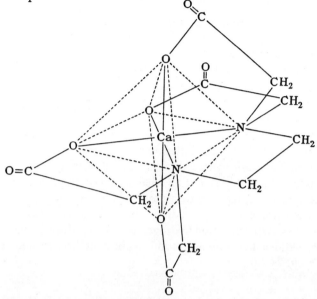

The complexes formed in most cases differ widely from the original ions. They are mostly soluble. The possibilities for the use of such reagents are extensive[2] and some of these may conveniently be reviewed under a number of headings:

(*i*) *Direct titration.* In the absence of a complexing cation ethylenediaminetetra-acetic acid acts with potassium hydroxide as a weak acid. In the presence of a complexing cation it acts as a strong acid, and may be titrated directly using ordinary acid-base indicators. As a consequence, the titration figure obtained against potassium hydroxide is a measure of the amount of the cation. Such cations as barium, lead, nickel and zinc have been titrated using this principle. In a typical example, 4 moles of potassium hydroxide are equivalent to 1 mole of EDTA and hence to one mole of cation.

(*i a*) *Indirect methods.* These offer a very fertile field of investigation. Thus, sodium may be precipitated as sodium zinc uranyl acetate, and the zinc then estimated by titration to give a measure of the sodium. Again, a silver salt in the presence of $Ni(CN)_4^-$ liberates its equivalent of nickel ions: consequently by a determination of nickel, silver, and through silver, chloride may be determined.

(*ii*) *Water hardness.* Calcium and magnesium may be titrated together using certain dyes as indicators, and magnesium only is titrated when using certain other dyes as indicators, thus giving calcium hardness by difference.

(*iia*) *Indirect methods.* Titrations may be effected, against the magnesium complex, of cations which produce more stable complexes, using suitable indicators. Again, through magnesium, magnesium ammonium phosphate and hence phosphate may be determined.

(*iii*) *Prevention of interference.* Many examples of this have been mentioned in the literature. Thus, 8-hydroxyquinoline precipitates only molybdenum, titanium, uranium, vanadium and tungsten in an acetate-buffered solution containing EDTA. As a consequence, molybdenum can be determined in the presence of interfering elements such as iron or aluminium. The oxine precipitates at other pH values are also selectively affected, so that, for example, iron can be determined in the presence of a considerable range of interfering elements.

The beryllium complex is not very stable, and beryllium hydroxide is precipitated by ammonia in the presence of EDTA. But aluminium and other interfering elements are retained in solution by EDTA.

In the gravimetric determination of sulphate and nitrate many ions which are normally adsorbed may be retained in solution, thus reducing interference from this cause.

Such reagents as sodium diethyldithiocarbamate and thioacetamide, which precipitate a wide range of cations, act much more selectively in the presence of EDTA.

(*iii a*) *Relative stabilities.* A cation forming a complex of low stability may be separated by adding a cation forming a complex of higher stability in the presence of an organic reagent and EDTA.

(*iv*) *Redox potentials.* The oxidation potentials of the complexes may be completely different from those of the simple ions. For example, cobalt-II can be oxidised to cobalt-III by ceric solutions, cobalt-III can be reduced to cobalt-II by chromous solution and manganese can be estimated oxidimetrically by ferricyanide.

(*v*) *Solution of "insoluble" precipitates.* Such "insoluble" materials as calcium oxalate and lead iodide are soluble in EDTA solutions.

(*vi*) *Colorimetry.* Certain cations, such as chromium, cobalt and manganese, form soluble coloured complexes with EDTA, and these may be used for quantitative determination of the cations.

In this section some uses of ethylenediaminetetra-acetic acid in determinations dependent on precipitation reactions will be described. Titrimetric uses of the reagent will be described in the appropriate section (Chapter VI).

Přibil and his co-workers[3] have studied the masking effect of nitrilo-triacetic acid and ethylenediaminetetra-acetic acid as masking reagents in a series of determinations. As a result of this work certain selective precipitations, hitherto impossible, can now be effected. For example, hydrogen sulphide, iodide and ammonia, as well as 8-hydroxyquinoline, are rendered more selective in the presence of these complexing agents.

When the complex formed by a tervalent metal with EDTA in a solution of ammonia is treated with a bivalent metal such as calcium, the tervalent metal hydroxide is precipitated in a pure state. Since the only tervalent metal in Group II A is bismuth, a separation as the hydroxide can be achieved in the presence of the other Group II A metals.[4] The method is rapid and accurate. In the presence of lead, chloride must be absent since some lead chloride may be precipitated. Bismuth may be precipitated in the presence of copper and cadmium without using EDTA since both copper and cadmium form soluble ammonia complexes. Consequently the addition of EDTA is of advantage only when lead is also present. Excellent results are obtained even in the presence of 800 mg. lead.

In the presence of EDTA, beryllium may be precipitated with ammonia in the presence of aluminium, bismuth, cadmium, chromium, cobalt, copper, iron, lead, manganese, nickel, zinc and small amounts of vanadium.[5] Titanium interferes and must be removed. When present in moderate amounts vanadium contaminates the precipitate and colours it yellow; presumably reprecipitation would be effective in removing the contaminant. Phosphate interferes by forming ammonium beryllium phosphate, but can be removed by precipitating with ammonium molybdate. The excess molybdate in the filtrate does not interfere. In the absence of other metals, aluminium or iron may be determined in the filtrate by precipitation with ammonia.

In the presence of EDTA, titanium can be precipitated quantitatively as hydroxide from a solution containing aluminium, bismuth, cadmium, iron, lead, manganese, mercury and nickel.[6] Chromium is coprecipitated when in the tervalent condition, and must first be oxidised to chromate in an alkaline medium. Beryllium interferes and should be absent. The precipitation must be effected in a cold solution, since at higher temperatures the precipitate is colloidal. The method can be applied in the analysis of bauxite.

Reagents

0·1 M Ethylenediaminetetra-acetic acid (di-ammonium salt). Dissolve 29·21 g. acid in a small excess of ammonia. Neutralise with nitric acid to phenolphthalein and dilute to 1000 ml. with distilled water.

0·5 M EDTA. Pour 40 ml. water on 29·21 g. reagent and add concentrated ammonia until alkaline to methyl red. Dilute to 200 ml. with water.

0·1 M Calcium nitrate. Dissolve 23·613 g. calcium nitrate, $Ca(NO_3)_2 . 4 H_2O$, in 1000 ml. water.

Ammonium nitrate wash solution. 1 per cent neutralised to methyl red with ammonia.

Procedures

Determination of bismuth. Add 5 ml. 0·1 M EDTA and 5 ml. 0·1 M calcium nitrate to the solution containing bismuth. Dilute with water, and add carbonate-free ammonia to the hot solution until the smell of ammonia persists. Allow the precipitate to settle, filter through an ashless filter paper, wash with hot water and ignite. Weigh as Bi_2O_3.

$$mg. \; ppt. \times 0·8970 \equiv mg. \; Bi.$$

If a large excess of calcium salt is used the precipitate is contaminated, and must be reprecipitated. Hence it is desirable only to add sufficient calcium salt to bind the complexone present.

Determination of bismuth in lead alloys. Weigh 1·0–25·0 g. alloy, depending on the bismuth content, and dissolve in nitric acid. If tin and antimony are absent precipitate directly with EDTA and an equivalent amount of calcium nitrate in the presence of ammonia, as already described. If tin is present, evaporate twice to precipitate *meta*stannic acid quantitatively. Filter, and extract the residue several times with hot EDTA solution. Determine the bismuth in the combined extracts and filtrate as already described. If antimony is present proceed as in the case of tin. However, some antimonic acid dissolves in the EDTA solution, and it is necessary to treat the

8

precipitate with sodium sulphide. The resulting bismuth sulphide is ignited to the oxide after filtration.

If tin is absent, and only traces of bismuth are present, dilute the solution to 400 ml., heat to boiling, and neutralise to methyl orange with ammonia solution. Add 4–5 g. pure calomel in suspension, and heat and stir for a short time. Allow the calomel to settle for 2–3 hours, filter, and wash thoroughly with water. Extract the coprecipitated bismuth hydroxide from the precipitate with hot 15 per cent nitric acid. Precipitate the bismuth from this solution and determine as already described.

Determination of beryllium. Neutralise 80–120 ml. of the solution containing about 50–80 mg. beryllium as oxide with ammonia solution until the hydroxides begin to precipitate. Redissolve by the addition of a few drops of dilute hydrochloric acid. Add 0·5 g. ammonium chloride and sufficient EDTA to complex all the heavy metals present (*e.g.*, 2 ml. 0·5 M EDTA per 27 mg. Al). Add 15–20 ml. 15 per cent ammonia solution to the cold solution whilst stirring continuously. Allow the precipitate to stand 2–3 hours or overnight, filter on an ashless filter paper, and wash with 100–150 ml. hot ammonium nitrate solution. If the aluminium is more than 130 per cent of the beryllium the precipitation must be repeated at this stage. Dissolve in dilute hydrochloric acid and continue as before. Ignite the paper and precipitate, completing the ignition over a blowpipe flame for 30 minutes. Weigh as BeO.

$$\text{mg. ppt.} \times 0\cdot3605 \equiv \text{mg. Be.}$$

Determination of aluminium or iron in the filtrate in absence of other metals. Evaporate the combined filtrates to about 80 ml. Add 15 ml. concentrated hydrochloric acid and 0·2–2·0 g. potassium chlorate according to the amount of EDTA present. Warm on the water bath until the odour of free chlorine disappears. Determine the aluminium or iron in the solution by precipitation with ammonia.

Determination of beryllium in the presence of chromium. It is necessary to warm the solution to form the stable purple complex of EDTA with tervalent chromium. Cool before precipitating beryllium. When the amount of chromium is four or more times that of beryllium reprecipitation is necessary because of adsorption.

Determination of beryllium in beryl and aquamarine. Separate silica by the conventional method from 0·6–1·0 g. finely ground sample. Dilute the filtrate to 250 ml. and take a suitable aliquot. Evaporate to small volume, partially neutralise with ammonia solution, and precipitate the beryllium as already described.

Determination of copper and beryllium in alloys. Dissolve 1·0–1·5 g. alloy in nitric acid and determine the copper electrolytically. Add ammonia solution until a permanent turbidity appears, clear with a few drops of dilute hydrochloric acid, and proceed as already described for beryllium.

Determination of titanium. Add to the acid solution containing titanium sufficient EDTA solution to complex the metals present, followed by ammonium chloride and ammonia. As soon as a turbidity appears stir vigorously, for example with an electric stirrer. After 5 hours' stirring filter on an ashless filter paper and wash with 2 per cent ammonium nitrate solution. Dry the precipitate, and ignite to constant weight. Weigh as TiO_2.

$$\text{mg. ppt.} \times 0·5995 \equiv \text{mg. Ti.}$$

Determination of iron in the filtrate when aluminium is absent. Destroy EDTA as described previously for the determination of aluminium or iron, and precipitate the iron as hydroxide in a volume not exceeding 60–80 ml. Alternatively, add an amount of calcium nitrate equivalent to the EDTA present prior to the precipitation with ammonia. Iron can be precipitated also with sodium hydroxide in the presence of EDTA, and separated from sodium salts with ammonia solution.

Determination of aluminium in the filtrate when iron is absent. Acidify the filtrate with hydrochloric acid and bind the EDTA by the addition of calcium nitrate solution. Add about 1 g. ammonium chloride and precipitate the aluminium as hydroxide with ammonia.

[1] G. Schwarzenbach, W. Biedermann and F. Bangerter, *Helv. Chim. Acta*, 1946, **29**, 811: G. Schwarzenbach and W. Biedermann, *ibid.*, 1948, **31**, 331, 459, 678.
[2] C. L. Wilson, *Ann. Reports*, 1951, **48**, 311; 1952, **49**, 304; 1953, **50**, 340.
[3] R. Přibil, *Coll. Czech. Chem. Comm.*, 1951, **16**, 86.
[4] R. Přibil and J. Čuta, *ibid.*, 391.
[5] R. Přibil and J. Kucharský, *ibid.*, 1950, **15**, 132.
[6] R. Přibil and P. Schneider, *ibid.*, 886.

11. 1 : 2-*cyclo*-HEXANEDIONEDIOXIME (NIOXIME)

This compound, which is soluble in water, reacts like dimethyl-glyoxime and other *anti*-dioximes to give a scarlet precipitate with nickel and a yellow precipitate with palladium.[1]

The reagent is a crystalline solid, m.p. 189°–190° C., which is quite white if free from contamination by iron. Its formula is

Its solubility in water is 0·82 g. per 100 ml. at 21·5° C., and it is used as a 0·8 per cent aqueous solution, which is stable indefinitely. A solution containing 1 part Ni^{++} in 10,000,000 shows a red colour after 2 minutes, and after standing 1 hour has deposited a red precipitate. A solution containing 1 part Pd^{++} in 2,000,000 shows a yellow colour after 5 minutes and deposits a yellow precipitate on standing for several hours. The gravimetric factors for these precipitates are rather more favourable than in the case of the corresponding complexes with dimethylglyoxime.

Quantitative precipitation of nickel occurs at pH 3·0 or higher. Both pH and rate of precipitation affect the crystalline form of the precipitate, and the best results are obtained by gradually raising the pH of the mixture of solution and precipitant from the point where the precipitate just begins to persist to pH 4·0 by slow addition of ammonium acetate solution. Above pH 7·0 the precipitate obtained is not satisfactory.[2]

Because of coprecipitation of the reagent, which is a linear function of the excess added, the volume of reagent solution must be measured approximately and an empirical correction applied. This is based on the fact that each 20 per cent excess of reagent increases the weight of the precipitate by 0·1 mg. per 25 mg. nickel present. The correction need only be applied for amounts of nickel above 15 mg. Amounts of nickel from 5 to 25 mg. can be successfully determined, but below 3 mg. precipitation is rather slow.

The precipitate may be dried at 100°–155° C. without decomposition.

Above 155°C. some decomposition occurs, and if ignition to the oxide is attempted, the precipitate sublimes.*

Alkali metals, alkaline earth metals, arsenic, beryllium, cadmium, manganese, uranium, zinc, acetate, chloride, nitrate, perchlorate, sulphate, sulphosalicylate and tartrate do not interfere. Aluminium and antimony are prevented from precipitating by addition of tartrate as a complexing agent. The determination fails in the presence of iron, since if not complexed, iron-III is partially reduced by the reagent to iron-II, which forms a particularly stable complex; and no complexing agent has been found which is capable of preventing interference of iron and at the same time allows normal precipitation of the nickel complex. Copper gives a coloration with the reagent, and cobalt coprecipitates.

Quantitative precipitation of palladium occurs in the pH range 0·7–3·0. Because of the low solubility of the complex it may be filtered from the hot solution after a brief digestion period, as opposed to the determination of this element using dimethylglyoxime, where a long period of standing is recommended. No coprecipitation of reagent is found to occur, even with 150 per cent excess reagent. The recommended excess is 30 per cent, *i.e.*, about 0·4 mg. reagent for each mg. palladium.

Amounts of palladium of 6 to 30 mg. can be determined successfully. Alkali metals, alkaline earth metals, aluminium, beryllium, cadmium, lanthanum, platinum, ruthenium, uranium, zinc, acetate, chloride, nitrate, sulphate, sulphosalicylate and tartrate do not interfere. Gold-I gives a precipitate.

Procedures

Determination of nickel. Adjust the volume of the solution, containing about 25 mg. nickel, to about 250 ml. Add any complexing agent necessitated by the presence of aluminium or antimony. From a graduated cylinder add 8 ml. 0·8 per cent aqueous reagent solution for each 10 mg. nickel expected. Redissolve any precipitate which forms by adding concentrated hydrochloric acid slowly with stirring. Add ammonia dropwise till a faint red coloration persists. Warm the solution to 60°C. and add 25 ml. 20 per cent ammonium acetate solution dropwise from a burette with constant stirring. Digest the solution for 30–40 minutes at 60°C., stirring occasionally. The pH

---

* C. Duval, *loc. cit.*, p. 234, states that the precipitate explodes at 115°C.

should now be 4·0–5·0. Collect the precipitate on an X3 sintered-glass filter crucible, wash five times with hot water, dry for 1 hour at 110° C., and weigh as $Ni(C_6H_9O_2N_2)_2$.

$$\text{mg. ppt.} \times 0\cdot1721 \equiv \text{mg. Ni.}$$

Where the amount of nickel is greater than 15 mg. the empirical correction recommended reduces to

$$\text{mg. ppt.} \times 0\cdot1721 \left[0\cdot98 - \frac{0\cdot00208\ (\text{mg. ppt.})}{(\text{ml. }0\cdot8\text{ per cent reagent added})} \right] \equiv \text{mg. Ni.}$$

Determination of palladium. Adjust the volume of the solution, containing 5–20 mg. palladium, to 200 ml. and the pH to any point from 1–5 depending on the other cations present. Warm the solution to 60° C. and add slowly from a pipette 0·43 ml. 0·8 per cent aqueous reagent solution for each mg. palladium present. Digest for 30 minutes at 60° C., stirring occasionally. Collect the precipitate on an X3 sintered-glass filter crucible, wash five times with hot water, dry at 110° C., and weigh as $Pd(C_6II_9O_2N_2)_2$.

$$\text{mg. ppt.} \times 0\cdot2743 \equiv \text{mg. Pd.}$$

[1] R. C. Voter, C. V. Banks and H. Diehl, *Analyt. Chem.*, 1948, **20**, 458.
[2] *Idem, ibid.*, 652.

12.   1 : 2-*cyclo*-HEPTANEDIONEDIOXIME (HEPTOXIME)

Of the water-soluble 1 : 2-dioximes which react with nickel, 1 : 2-*cyclo*-hexanedionedioxime (Nioxime p. 100) and 1 : 2-*cyclo*-heptanedionedioxime (Heptoxime) are sufficiently soluble in water to be used in aqueous solution at room temperature as precipitants. Heptoxime possesses certain advantages over Nioxime as a gravimetric reagent.[1] The reagent has the formula

It is a white crystalline solid which can be recrystallised from water, being soluble in that solvent to an extent of 4·8 g. per l. at 19·5°C. It is therefore less soluble than Nioxime (8·2 g. per l. at 21·5°C.) but very much more soluble than dimethylglyoxime (0·4 g. per l. at room temperature). As obtained by recrystallisation from water, the compound contains one molecule of water of crystallisation, which is lost at 82°–87°C. Small amounts may be recrystallised from benzene to give the anhydrous compounds, m.p. 182°C.

Two molecules of Heptoxime are required for precipitation of one molecule of the complex nickel-Heptoxime compound, which is presumed to have the familiar co-planar structure of the other 1 : 2-dioxime complexes. However, the Heptoxime complex is yellow, and unlike nickel dimethylglyoxime is best precipitated from a slightly acid medium, precipitation being quantitative at pH 2·7 or greater. This permits of separation of nickel from certain other cations without using complexing agents to avoid interference. The most satisfactory pH for the precipitation is in the range of 3·5–4·0, and if citrate or tartrate are not used for complexing purposes ammonium acetate is used as a buffer. The presence of ammonium acetate also appears to reduce adsorption of the reagent by the precipitate.

The precipitate has a very satisfactory nature for the usual gravimetric operations, and does not creep as does the nickel dimethylglyoxime precipitate. It is quite stable, and may be dried for 5 hours at 120°C. without any decomposition.

Aluminium, arsenic, bismuth, antimony, chromium, iron-III and titanium interfere and must be complexed with tartrate or citrate. Cobalt reacts to form a fairly soluble brown complex. In the presence of cobalt, therefore, additional reagent will be necessary to complex the cobalt, and if considerable amounts are present a second precipitation is advisable. Interference by copper, which gives a brown insoluble precipitate, may be avoided by adding sufficient excess thiocyanate to the reduced solution to redissolve precipitated cuprous thiocyanate, which will not then react. If thiocyanate is used for this purpose the pH must first be adjusted to 3·5 to prevent formation of unstable perthionic acid. Lead, if present, would be precipitated as chloride, and this is prevented by addition of acetate. The common anions, and other cations, do not interfere.

The method may also be used on the micro scale.[2]

Procedure

Determination of nickel in iron and steel. Weigh the sample, which should contain about 20 mg. nickel for the macro determination, into a 500-ml. conical flask, and dissolve in a suitable acid medium. Decompose carbides with 10 ml. concentrated nitric acid, and add 10 ml. perchloric acid per gram of sample. As soon as the mixture begins to fume, boil for 15 minutes, add four times the volume of water, and filter off silica, washing the residue with 0·1 N hydrochloric acid and then with water. To the filtrate add 18 ml. 30 per cent citric acid solution for each gram of sample, and then 3 ml. in excess. Add 5 ml. 20 per cent ammonium acetate solution if lead is present (or if citrate or tartrate has not been used for complexing) and follow this by 10 ml. 10 per cent sodium sulphite solution. Dilute to 200 ml. with water, and adjust the pH to about 3·5 with ammonia. Warm to 50°C., add 20–30 ml. 50 per cent ammonium thiocyanate solution, and stir till cuprous thiocyanate redissolves.

To the clear solution add slowly with constant stirring 30 ml. saturated aqueous reagent solution for each 20 mg. nickel present, and then add 5 ml. in excess. Digest for 10 minutes at 80°C., and then place the vessel for 30 minutes in water to cool. Filter through an X3 sintered-glass filter crucible, keeping the filter filled with liquid. Wash the precipitate with cold water, and dry for 1 hour at 110°–120°C. Weigh as $Ni(C_7 H_{11}O_2N_2)_2$.

$$mg. \ ppt. \times 0 \cdot 1589 \equiv mg. \ Ni.$$

[1] R. C. Voter and C. V. Banks, *Analyt. Chem.*, 1949, **21**, 1320.
[2] R. C. Ferguson, R. C. Voter and C. V. Banks, *Mikrochem. Mikrochim. Acta*, 1951, **38**, 11.

13.   *tris*-(HYDROXYMETHYL)-AMINOMETHANE

A primary standard for the standardisation of dilute solutions of strong acids must have a high equivalent weight, be easy to obtain and to maintain in a high state of purity, and be non-hygroscopic and colourless in solution. *tris*-(Hydroxymethyl)-aminomethane, m.p. 171·1°C., has been proposed for this purpose.[1]

$$CH_2.OH$$
$$|$$
$$HO.H_2C- C -CH_2.OH$$
$$|$$
$$NH_2$$

This compound is easy to purify, and can be dried by heating at 100°–103° C., although it cannot be heated much above this temperature, decomposition being observed on heating to 110° C. It is no more hygroscopic than potassium hydrogen phthalate, and neither it nor its solutions absorb carbon dioxide from the atmosphere. Its solutions are stable.

The pH at its equivalence-point is 4·7, and *p*-sulpho-*o*-methoxybenzeneazodimethyl-1-naphthylamine is recommended as the most suitable indicator for use with it. However, ethyl orange or a mixed bromocresol green-sodium alizarin sulphonate indicator are also satisfactory. The equivalent of the compound is 121·12.

Reagents

p-*Sulpho*-o-*methoxybenzeneazodimethyl*-1-*naphthylamine*. Triturate 0·1 g. of indicator in a mortar with 2·59 ml. 0·1 N sodium hydroxide solution, and dilute to 100 ml. with boiled distilled water.

Ethyl orange. Dissolve 0·1 g. in 100 ml. boiled distilled water.

Mixed indicator. Triturate 0·1 g. bromocresol green with 1·45 ml. 0·1 N sodium hydroxide solution and dilute to 100 ml. with boiled distilled water. Dissolve 0·1 g. sodium alizarin sulphonate (alizarin red S) in 100 ml. boiled distilled water. Mix equal volumes of these two solutions.

Reference buffer solution: (pH 4·7). Mix 50 ml. 0·1 N potassium hydrogen phthalate solution, 13·1 ml. 0·1 N sodium hydroxide solution, 3 drops indicator solution (6 drops of the mixed indicator) and 25 ml. distilled water.

Procedure

Standardisation of acid. Weigh accurately about 0·5 g. of the standard into a 250-ml. tall-form beaker. Dissolve in 50 ml. freshly boiled and cooled distilled water. Add the appropriate amount of indicator solution, and titrate with acid to the end-point, which may be determined by matching with the colour of the indicator in the reference buffer solution.

Purification of the standard

Dissolve 1200 g. *tris*-(hydroxymethyl)-aminomethane in 2400 ml. distilled water at 60° C. in a 4-litre beaker. Add 50 g. decolorising charcoal and keep at 50°–60° C. with constant stirring for 30 minutes, filter hot through a coarse sinter carrying a 0·25-inch pad of paper pulp, and collect in a 4-litre suction flask. Repeat this process. Transfer the solution to a 4-litre beaker, add a few glass beads, cover with a clock glass, and boil slowly to concentrate the solution to a volume of about 1300 ml.

Divide this solution into two portions and pour each of them down the side of a 4-litre beaker containing 2000 ml. pure redistilled methanol. Continue to stir while recrystallisation is taking place. Allow to cool for about 30 minutes, and immerse the beakers in an ice-salt slurry, to cool to 3°–4°C. Filter off the liquid through a coarse sintered-glass filter. Wash the solid by slurrying once with cold methanol, and dry by suction. Combine the two portions.

Dissolve the combined product in 1000 ml. water distilled from alkaline permanganate, concentrate to a total volume of 1000 ml., and pour into 3000 ml. methanol, separating as before. Redissolve in 800 ml. water distilled from alkaline permanganate, concentrate to a total volume of 800 ml., and pour into 2500 ml. methanol. Separate as before. Wash the product in the funnel by slurrying twice with cold methanol, dry by rapid suction and then by pressing between sheets of filter paper. Grind the product to a fine powder, place in open dishes, and dry in a vacuum desiccator over phosphorus pentoxide for 24–36 hours at 5 mm. pressure.

[1] J. H. Fossum, P. C. Markunas and J. C. Riddick, *Analyt. Chem.*, 1951, **23**, 491.

14.   2-(o-HYDROXYPHENYL)-BENZIMIDAZOLE

Walter and Freiser[1] have reported that 2-(o-hydroxyphenyl)-benzimidazole

is a much more selective reagent for mercury than reagents such as mercaptobenzimidazole, tetraphenylarsonium chloride, azo derivatives of 8-hydroxyquinoline or thionalide, the reagents most generally used for this element. It can be applied to either the gravimetric or the titrimetric determination of mercury in the presence of a large range of elements which normally interfere with the determination.

The reagent is used in ethanolic solution, and forms a yellow precipitate with a composition corresponding to $(C_{13}H_9N_2O)_2Hg$. This precipitate is stable to quite high temperatures, darkening only being observed at 250° C. The precipitate can conveniently be dried at 130°–140° C. It is insoluble, or only very slightly soluble, in a wide range of organic solvents, including ethanol, which can therefore be used in washing the precipitate free from reagent. Although only a small excess (6–10 per cent) of reagent is required for complete precipitation, the precipitate may be washed free from up to 90 per cent excess with 50 per cent aqueous ethanol.

Precipitation is complete in the range pH 6·0–7·0, and citrate is used to complex elements which would precipitate as hydroxides in this range. No precipitate is formed with aluminium, arsenic, barium, bismuth, cadmium, chromium, cobalt, copper, iron-II, iron-III, lead, magnesium, manganese, nickel, potassium, silver, sodium, tin-II or zinc, although colour changes with some of these metals suggest the formation of soluble complexes. In the range 10–100 mg. mercury, precipitation of mercury is found to be quantitative in the presence of any of these elements excepting iron-II and iron-III, giving a gravimetric accuracy of better than ±0·3 mg. Both iron-II and iron-III interfere seriously if present in anything other than very small amounts. Attempts to mask the interfering action of iron by complexing were unsuccessful.

Since the reagent is a phenol it may be brominated quantitatively, permitting the determination of mercury titrimetrically by dissolving the complex in acetic acid and bromination by the usual methods. This reaction is not, however, a simple bromination one, since each molecule of reagent consumes three molecules of bromine, one of which is used in brominating and the other two in oxidation. The reactions taking place are as follows:

$$Hg^{++} + 2C_{13}H_{10}ON_2 \longrightarrow (C_{13}H_9ON_2)_2Hg + 2H^+.$$

$$(C_{13}H_9ON_2)_2Hg + 2CH_3COOH \longrightarrow Hg^{++} + 2CH_3COO^- + 2C_{13}H_{10}ON_2.$$

$$2C_{13}H_{10}ON_2 + 6Br_2 \longrightarrow 2C_{13}H_6ON_2Br_2 + 8HBr.$$

It can be seen, therefore, that each atom of mercury is equivalent to six molecules of bromine, so that the titrimetric determination is highly sensitive. Using this procedure the accuracy in the 80 mg. range is ± 0.2 mg. The principal disadvantage is that the precipitated bromo compound colours the solution in the same way as iodine, so that the end-point cannot be detected visually and electrometric methods must be used.

Solution required

Reagent solution. Dissolve 1 g. in 100 ml. 95 per cent ethanol.

Procedures

Gravimetric determination of mercury. To the solution, containing 10–100 mg. mercury and free from all but traces of iron, add 2 g. sodium citrate. Warm to 60° C. and adjust to pH 5 with dilute acetic acid or sodium hydroxide. Add sufficient reagent solution to precipitate the mercury completely and leave a small excess. Raise the pH to 6·5 with 0·3 N sodium hydroxide solution. Maintain the solution and precipitate at 60° C. for about 15 minutes, and allow to cool to room temperature.

Filter through an X3 sintered-glass filter crucible, wash three or four times with 15 ml. 50 per cent ethanol, and dry at 130°–140° C. to constant weight.

$$\text{mg. ppt.} \times 0.3240 \equiv \text{mg. Hg.}$$

Titrimetric determination of mercury. Precipitate the complex, containing about 80 mg. mercury, as already described, and wash the precipitate thoroughly with 50 per cent ethanol. Dissolve in 50 ml. hot glacial acetic acid, transfer the solution to an iodine flask, and dilute slightly with distilled water. Add 60 ml. 0·1 N potassium bromate-potassium bromide solution, stopper the flask immediately, and add a few ml. potassium iodide solution to the reservoir of the iodine flask to avoid loss of bromine. Immerse the flask in a beaker of water and maintain at 35° C. for from one to one and a quarter hours. Add a solution containing 1·5 g. potassium iodide, and titrate the liberated iodine with 0·1 N sodium thiosulphate solution, observing the end-point with a dead-stop indicator apparatus.[2]

$$\text{1 ml. 0·1 N potassium bromate} \equiv 1.672 \text{ mg. Hg.}$$

Synthesis of the reagent

Mix 1 mole (108 g.) *o*-phenylenediamine and 1 mole (137 g.) salicylamide in a mortar until finely powdered; transfer to a round-bottomed flask and heat in an oil-bath at 160° C. until the mixture has melted completely and ammonia and water vapour are evolved. Raise the temperature to 200°C. and maintain at this temperature until ammonia ceases to be evolved.

Transfer the molten mixture to a distilling flask, and distil completely at ordinary pressure. Recrystallise the yellow solid obtained three times from an ethanol-water mixture.

Yield 30 per cent. M.p. 242°C.

[1] J. L. Walter and H. Freiser, *Analyt. Chem.*, 1953, **25**, 127.
[2] H. H. Willard, L. L. Merritt and J. A. Dean, *Instrumental Methods of Analysis*, 2nd Ed., New York, 1951, p. 212.

15. 2-(o-HYDROXYPHENYL)-BENZOXAZOLE

The gravimetric determination of cadmium by precipitation with 2-(o-hydroxyphenyl)-benzoxazole has been described by Walter and Freiser.[1] The reagent has the formula

and produces with cadmium in alkaline solution a complex which corresponds to the formula $Cd(C_{13}H_8O_2N)_2$. In addition to being insoluble in water, the complex is insoluble in many organic solvents. It is stable at least to 275°C., and does not melt even at 300°C. although it has begun to darken at this temperature. Only a slight excess of reagent (5–10 per cent) should be used, although the precipitate may be freed from 75 per cent excess by washing with 50 per cent ethanol.

The only common ions which interfere are cobalt, copper and nickel. Copper precipitates at pH 3·5–4·0, and therefore amounts up to 100 mg. may be removed by precipitation in acid solution. Cadmium does not begin to precipitate in the presence of tartrate until pH 6·5, and is best precipitated at pH 11·0–12·0. Calcium, if present, is then precipitated as calcium tartrate, and is filtered off, sufficient additional tartrate being added to the filtrate to compensate for the amount precipitated. With considerable excess of tartrate other metallic ions do not precipitate, with the exception of cobalt and nickel in amounts greater than 20 mg. If amounts of these ions in excess of 20 mg. are present, they are first removed by treatment with 1-nitroso-2-naphthol and dimethylglyoxime respectively.

Solution required

Reagent solution. Dissolve 1 g. reagent in 100 ml. 95 per cent ethanol.

Procedure

Determination of cadmium. Add 3 g. ammonium tartrate to the solution containing 1–80 mg. cadmium. Follow this by sufficient 3 N ammonium acetate to bring the pH to about 3·5 Warm the solution to about 60° C. and filter off any calcium tartrate precipitate. If such a precipitate forms, add a further 1 g. ammonium tartrate to the filtrate.

Add sufficient freshly prepared reagent solution to give a slight excess over the amount required for complete precipitation of the copper, cobalt and nickel, and add 3 N sodium acetate solution to bring the pH to 4·0. Digest at 60° C. for 15 minutes, allow to cool to room temperature, and filter through an X3 sintered-glass filter crucible. Wash the precipitate several times with water, combining the washings with the filtrate. Bring the pH of the solution to approximately 9·0 with N sodium hydroxide, heat once more to 60° C., and add a slight excess of reagent solution. Add sufficient N sodium hydroxide solution to bring the pH to 11·0, digest at 60° C. for 15 minutes, and allow to cool to room temperature. Filter through a weighed X3 sintered-glass filter crucible, wash several times with ammoniacal 50 per cent ethanol, and dry for 2 hours at 130°–140° C. Weigh as $Cd(C_{13}H_8O_2N)_2$.

$$mg. \ ppt. \times 0·2109 \equiv mg. \ Cd.$$

Synthesis of the reagent[2]

Mix 1 mole o-aminophenol and 1 mole salicylamide in a mortar until finely powdered, transfer to a round-bottomed flask, and heat in an oil-bath till melted (about 90° C.) when ammonia and water vapour are evolved. After 1 hour raise the temperature to 200° C. and maintain at this temperature till no further ammonia is produced (about 3 hours). Transfer the melted mixture to a distilling flask and distil over the red 2-(o-hydroxyphenyl)-benzoxazole at 338°–340° C. Recrystallise three times from ethanol. Yield about 48 per cent. M.p. 124° C.

[1] J. L. Walter and H. Freiser, *Analyt. Chem.*, 1952, **24**, 984.
[2] S. Siegfried and M. Moser, *Ber.*, 1922, **55**, 1089.

16. 8-HYDROXYQUINALDINE

8-Hydroxyquinaldine, or 2-methyl-8-hydroxyquinoline,

one of the simplest 2-substituted derivatives of oxine, has been found[1] to be more selective than the parent compound. In acetic acid-acetate solutions, bismuth, cadmium, chromium, cobalt, copper, iron-II, iron-III, manganese, nickel, silver, titanium-IV, zinc, molybdate, tungstate and vanadate form insoluble complexes. Aluminium, ammonium, barium, beryllium, calcium, lead, magnesium, potassium and sodium do not precipitate. If tartrate is present, bismuth and tin-IV remain in solution.

In ammoniacal solution containing tartrate to prevent precipitation of aluminium hydroxide, the ions which form insoluble complexes in the buffered acetic acid solution, with the exception of molybdate, tungstate and vanadate, and the addition of calcium, lead, magnesium and strontium, are precipitated.

It is important to note that aluminium in particular, which is readily determined using 8-hydroxyquinoline, does not precipitate with 8-hydroxyquinaldine. Consequently it is possible, using this reagent, to determine zinc, which does precipitate, in the presence of aluminium. The determination may be completed gravimetrically by weighing the precipitated complex, or titrimetrically by bromination. The aluminium may subsequently be determined, after removal of zinc in this way, by 8-hydroxyquinoline.

Zinc can also be determined in the presence of magnesium, since the zinc complex precipitates in acetic acid-acetate solutions (pH 5·5) while magnesium precipitates only at a pH of 9·3 or higher.

If zinc, magnesium and aluminium are present together the magnesium determination fails, since tartrates and ammonium salts must be added to maintain the aluminium in solution, and in these conditions the magnesium complex precipitates too slowly to be of practical analytical use.

Calcium ion in amounts above 2–3 mg. interferes with the magnesium determination, and should first be removed.

Both the reagent and the complex are more soluble in water containing ethanol than in water alone. Consequently when using an ethanolic solution of the reagent only the required amount, as shown by a yellow colour in the filtrate, should be added, and much excess should be avoided. In alkaline solution the reagent does not coprecipitate, and any reagent which coprecipitates in acid solution is volatile at 130° C.

As an alternative to a bromometric titration, determination of the magnesium complex by reaction with potassium permanganate has been proposed.[2] When the 8-hydroxyquinoline complexes of metals are redissolved in sulphuric acid and excess of standard potassium permanganate solution is added, the permanganate solution reacts slowly with the free organic reagent. The excess of permanganate may then be reduced by potassium iodide, and the liberated iodine determined by standard thiosulphate solution. This reaction is not, however, stoicheiometric, and accurate results are only obtained if an empirically established procedure is rigidly followed.

On the other hand, 8-hydroxyquinaldine appears to react stoicheiometrically with the permanganate according to the equation

$$2C_9H_7ON + 6KMnO_4 + 9H_2SO_4 \longrightarrow 2C_5H_3N(COOH)_2$$
$$+ 3K_2SO_4 + 6MnSO_4 + 3CO_2 + H.COOH + 10H_2O$$

and the titration may then be finished iodometrically.

Solutions required

Reagent solution. Dissolve 5 g. 8-hydroxyquinaldine in 12 ml. glacial acetic acid, and dilute to 100 ml. with water. This reagent solution is stable for at least 1 week.

Ammonium acetate solution. 154 g. per l.

Procedures

Gravimetric determination of zinc in absence of aluminium. To the sample solution add dilute ammonium hydroxide until a permanent precipitate of zinc hydroxide appears. Redissolve this with a drop of acetic acid. Add 2 ml. reagent solution to the sample solution for each 10 mg. zinc present, and then add 2–3 drops 15 N ammonium hydroxide solution. Dilute to about 200 ml. with water, and heat to 60°–80°C. Neutralise excess acid by adding 3 N ammonium hydroxide solution until the precipitate formed as each drop enters the solution only just redissolves on stirring. Add slowly 45 ml. ammonium acetate solution, stirring continuously. Set aside for 10–20 minutes. For low amounts of zinc set aside for several hours.

Filter through a tared Gooch or porous-porcelain filter crucible. Wash the precipitate well with hot water, dry at 130°–140°C. for at least 2 hours, and weigh as $Zn(C_{10}H_3ON)_2$.

$$mg.\ ppt. \times 0.1713 \equiv mg.\ Zn.$$

Gravimetric determination of zinc in presence of aluminium. Add 1 g. ammonium tartrate to the slightly acid solution. Add 2 ml. reagent solution for every 10 mg. zinc present. Complete the determination as before.

Titrimetric determination of zinc. Precipitate the complex by the appropriate method, and after waiting for the required time filter through a filter paper. Wash the precipitate well with hot water, dissolve the precipitate through the filter with 30 ml. hot 4 N hydrochloric acid, and wash the filter thoroughly with hot 3 N hydrochloric acid. Moisten the filter paper with a few drops concentrated hydrochloric acid, and wash with hot water. Repeat this procedure a second time. When the amount of zinc is small and the amount of magnesium or aluminium large, reprecipitate the zinc with 1–2 ml. reagent solution and redissolve.

Add 3 g. solid potassium bromide to the solution, dilute to about 150 ml., and add a few drops methyl red indicator solution. Run in 0.1 N potassium bromate solution until the indicator bleaches, and then add 5 ml. in excess. Add 3 g. solid potassium iodide, stir until dissolved, and back titrate with standard 0.05 N sodium thiosulphate solution, using 2 per cent starch solution as indicator. Standardise the bromate solution against arsenious oxide, using methyl orange

9

as indicator, and the sodium thiosulphate solution against the potassium bromate solution.

1 ml. 0·1 N potassium bromate ≡ 0·8175 mg. Zn.

Gravimetric determination of magnesium. To the filtrate from the zinc determination add 3 ml. reagent solution for every 10 mg. magnesium present, followed by concentrated ammonia solution until no further precipitate forms. Digest for 20 minutes at 60°–80° C. Filter through a tared Gooch or porous-porcelain crucible. Wash with hot water, and dry for 2 hours at 130°–140° C. Weigh as $Mg(C_{10}H_8ON)_2$.

mg. ppt. × 0·0714 ≡ mg. Mg.

Titrimetric determination of magnesium using potassium bromate. To the filtrate from the zinc determination add 3 ml. 5 per cent reagent in 95 per cent ethanol for each 10 mg. magnesium present. Add concentrated ammonia until no further precipitate forms. Filter through a filter paper and wash with hot water.

Dissolve the precipitate in hydrochloric acid in the same manner as described for the solution of the zinc complex, and complete the determination with potassium bromate as before.

1 ml. 0·1 N potassium bromate ≡ 0·3040 mg. Mg.

Titrimetric determination of magnesium using potassium permanganate. Precipitate the magnesium (0·01–0·05 g.) in the normal fashion, wash the precipitate, and dissolve it in 10 per cent sulphuric acid, making the volume up to 250 ml. Measure out 25 ml. of this solution, and add a measured excess of 0·1 N potassium permanganate solution. Set aside for 15 minutes or longer. Add 0·5 g. potassium iodide, and determine the liberated iodine in the usual way by titration with 0·1 N sodium thiosulphate solution.

1 ml. 0·1 N potassium permanganate ≡ 0·0811 mg. Mg.

Gravimetric determination of aluminium following determination of zinc. Warm the filtrate from the zinc determination to 60°–80° C., and add 40 ml. 2·5 per cent 8-hydroxyquinoline in 7·5 per cent acetic acid followed by 10 ml. ammonium acetate solution. Digest for 10–20 minutes, and filter through a tared Gooch or porous-porcelain filter crucible. Wash the precipitate thoroughly with hot water, and dry for at least 2 hours at 130°–140° C.

mg. ppt. × 0·0585 ≡ mg. Al.

Titrimetric determination of aluminium. Precipitate the 8-hydroxyquinoline complex as for the gravimetric determination, and filter through a filter paper. Wash thoroughly with hot water, and dissolve in hydrochloric acid, completing the determination as described for zinc.

1 ml. 0·1 N potassium bromate≡0·2249 mg. Al.

Synthesis of 8-hydroxyquinaldine

Dissolve 55 g. o-aminophenol and 25 g. o-nitrophenol in 100 g. 12 N hydrochloric acid in a three-necked flask fitted with a reflux condenser, a mechanical stirrer and a dropping funnel. Add 40 g. crotonaldehyde with stirring, over a period of 45 minutes. Continue stirring while heating on the steam-bath for 6 hours. Set aside overnight.

Steam distil to remove excess o-nitrophenol. Nearly neutralise the residual solution with 6 N sodium hydroxide solution. Saturate with solid sodium carbonate, and steam distil. Yield of crude 8-hydroxyquinaldine 24–32 g. (30–40 per cent).

Distil the crude material under reduced pressure in a sublimation tube, and recrystallise the light red product from 60 per cent aqueous ethanol. The resulting material is faintly yellow. Yield 20–26 g. (24–33 per cent). M.p. 69°C.

This material is pure enough for analytical purposes, but may be recrystallised once again, to give a product with m.p. 72°–74°C.

[1] L. L. Merritt and J. K. Walker, *Ind. Eng. Chem. Anal.*, 1944, **16**, 387.
[2] J. P. Phillips and F. J. O'Hara, *Analyt. Chem.*, 1951, **23**, 535.

17. 7-IODO-8-HYDROXYQUINOLINE-5-SULPHONIC ACID

Ferron, or 7-iodo-8-hydroxyquinoline-5-sulphonic acid,

better known as a colorimetric reagent for iron, forms a complex with calcium ions having the composition:

$$Ca(C_9H_4N.I.SO_3H.O)_2 . \tfrac{10}{3} H_2O.$$

This bright orange-red precipitate is formed in neutral or acid solutions, and its composition varies very slightly with the calcium content of the test solution. It is, however, satisfactory for the gravimetric determination of small amounts of calcium,[1] and it is possible to determine as little as 5 mg. calcium per litre. Magnesium, potassium or sodium ions do not interfere.

The separation of thorium from uranium and its subsequent determination takes less time using 7-iodo-8-hydroxyquinoline-5-sulphonic acid as precipitant than by the oxalate method. Under specified conditions cerium, cobalt, lanthanum, nickel and titanium do not interfere. Copper, mercury and silver precipitate under the same conditions as thorium.

When uranium is present it is difficult to judge the amount of reagent necessary for complete precipitation because of the formation of a coloured uranium complex. Complete precipitation of the thorium is achieved by addition of a considerable excess of reagent solution. The separation of thorium from twice its amount of uranium is satisfactory.

If the uranium content exceeds twice the thorium content a reprecipitation procedure may be used, but recoveries are not complete. Because of formation of interfering substances, wet oxidation methods cannot be used to prepare the precipitate for reprecipitation, and a careful combustion technique must be employed.[2] If the ignition is carried out at too low a temperature, sulphur is not volatilised from the complex, and sulphate is formed on later addition of nitric acid, interfering with the determination. If the ignition temperature is too high the resulting thorium oxide is too difficult to redissolve.

Solutions required

Ferron solution. 0·2 per cent aqueous solution.

Alkaline Ferron solution. Shake 8·8 g. Ferron with 200 ml. water and 6·5 ml. 4 N sodium hydroxide solution. Dilute to 250 ml. with water and filter.

Buffer solution. Add 10 g. crystalline sodium acetate to 80 ml. 30 per cent acetic acid. Dilute to 100 ml. with water. The pH of this solution is 4·0.

Procedures

Gravimetric determination of calcium. Evaporate the neutral or acid solution, containing not more than 50 mg. calcium, to a volume of 2·5 ml. To this add 2·5 ml. buffer solution, and heat on a water-bath. Add 25 ml. boiling alkaline Ferron solution, and shake the mixture thoroughly. Heat for a further 15 minutes on the water-bath, cool, and set aside overnight. Filter, and wash the precipitate, first with the alkaline reagent solution, then with ethanol and finally with acetone. Dry below 80° C. and weigh.

$$\text{mg. ppt.} \times 0.0501 \equiv \text{mg. Ca.}$$

Gravimetric determination of thorium. Transfer 10 ml. thorium solution to a 100-ml. beaker and adjust the pH to 2·0–3·5 by the addition of 10 per cent ammonium acetate solution and hydrochloric acid. Place on the steam bath for 10 minutes, and add Ferron solution gradually until the first permanent precipitate is obtained. Add a further 25 ml. Ferron solution. Digest for 30 minutes, filter through a No. 42 Whatman filter paper, wash with Ferron solution, char the paper over a burner, and ignite to the dioxide.

$$\text{mg. ppt.} \times 0.8788 \equiv \text{mg. Th.}$$

Separation of thorium from larger amounts of uranium. Produce the precipitate as already described, and filter through a porous-porcelain filter crucible. Heat for 1 hour at 550°–600° C. Wash the residue into the original beaker with dilute nitric acid, and evaporate to dryness. Evaporate three times further, first with 10 ml. of concentrated hydrochloric acid and 5 ml. concentrated nitric acid, then with 5 ml. concentrated nitric acid, and finally with water. Add 10 ml. water, and reprecipitate as described previously.

[1] N. Schoorl, *Pharm. Weekblad*, 1939, **76**, 620.
[2] D. E. Ryan, W. J. McDonnell and F. E. Beamish, *Ind. Eng. Chem. Anal.*, 1947, **19**, 416.

18. MANDELIC ACID AND ITS DERIVATIVES

A considerable number of interferences from other metals affect the precipitation of zirconium by the usual organic reagents. Kumins[1] found that with mandelic acid, $C_6H_5.CHOH.COOH$, there is no interference from aluminium, antimony, barium, bismuth, cadmium, calcium, cerium, chromium, copper, iron, thorium, tin, titanium or vanadium. The precipitation is effected in a hydrochloric acid medium. Sulphuric acid may be used, but the results are low when more than 5 per cent of free sulphuric acid is present. Sodium hydroxide may be used to reduce high acidity arising from free sulphuric acid, but not ammonia, which gives low results, probably because of the formation of a complex ammonium zirconium mandelate.

The precipitate corresponds to the formula

$$(C_6H_5.CHOH.COO)_4Zr,$$

and Feigl[2] considers that the structure is probably

Hahn[3] examined Kumin's procedure and found that much smaller amounts than those originally claimed could be determined accurately. The lowest quantity he determined was of the order of 0·1 mg. He also showed that the concentration of hydrochloric acid was not critical, and could range from 0·1 to 8 M. The effect on the determination of the presence of cobalt, magnesium, manganese, mercury, nickel, uranium and zinc, which had not been investigated by Kumins, was also investigated, and it was found that none of these elements interfered. Hafnium, however, is precipitated along with zirconium. A determination may be completed in less than an hour.

Oesper and Klingenberg[4] studied the behaviour of the following compounds related to mandelic acid in respect of their zirconium-precipitating capacity:

(1) p-Methylmandelic acid.
(2) p-Nitromandelic acid.
(3) p-Bromomandelic acid.
(4) p-Chloromandelic acid.
(5) p-Iodomandelic acid.
(6) o-Hydroxydecanoic acid.
(7) Benzilic acid.

(8) 2-Naphthylglycollic acid.
(9) o-Nitromandelic acid.
(10) m-Nitromandelic acid.
(11) Lactic acid.
(12) Glycollic acid.

All these related compounds give precipitates with zirconium under suitable conditions, but (9) to (12) are much less sensitive than mandelic acid. Large weighing effects are exhibited by (5), (7) and (8), but the reagents are water-insoluble, so that complications would arise in their use. No particular advantage over mandelic acid is shown by (1). Only (3) and (4), which show greater sensitivity than mandelic acid, were subjected to further investigation.

The precipitates with these compounds are completely insoluble in water, and may be washed with pure water without loss, as distinct from zirconium mandelate, for which a special wash solution is required. Moreover, the precipitates obtained using both substituted mandelic acids may be weighed as such, whereas zirconium mandelate must be ignited to the oxide, not because of variability of composition, but because a high concentration of reagent must be used, and it is difficult to remove this by washing because of the significant solubility of zirconium mandelate.

Although satisfactory results are obtained by direct weighing of the precipitate from the substituted mandelic acids when the method is tested on pure zirconium solution, erratic results are obtained when the procedure is followed on ore and alloy samples. Hence the ignition method, which gives satisfactory results on such samples, is still preferred.

In other respects the reagents behave like mandelic acid itself. Free sulphuric acid should not be present in amounts greater than 5 per cent. Sodium hydroxide, but not ammonia, may be used for neutralising any excess over this amount. The selectivity of the reagents is the same. The order of effectiveness of the reagents is stated to be p-bromomandelic acid, p-chloromandelic acid and mandelic acid.

Procedures

Determination of zirconium with mandelic acid. Adjust the volume of the solution, containing up to 300 mg. zirconium, to about 20 ml. with concentrated hydrochloric acid. Add 50 ml. 16 per cent mandelic acid solution, and dilute to 100 ml. with water. Heat slowly to 85°C. and maintain at this temperature for 20 minutes. Filter, and wash with a hot solution containing 2 per cent hydrochloric acid and 5 per cent mandelic acid. Ignite, and weigh as ZrO_2.

$$mg. \ ppt. \times 0·7403 \equiv mg. \ Zr.$$

Determination of zirconium in zirconium chloride solutions. Heat 50 ml. of the solution containing 15 ml. concentrated hydrochloric acid to 85°C. Add slowly from a burette 50 ml. 0·1 M solution of *p*-bromomandelic or *p*-chloromandelic acid. Digest the solution at 85°C. for 15–20 minutes. Filter on a porous-porcelain filter crucible, wash with distilled water, ignite, and weigh as ZrO_2.

Determination of zirconium in alloys. Treat 0·15 g. alloy in a 150-ml. beaker with 5 ml. water, 2 ml. concentrated sulphuric acid and 3 drops (or more if necessary) hydrofluoric acid. Cover the beaker with a watch glass. Towards the end of the reaction add 1 ml. nitric acid. Evaporate to dryness on a sand bath.

Dissolve the residue in 50 ml. water containing 1 ml. concentrated hydrochloric acid. Add the reagent as described previously, and finish the determination in exactly the same way.

[1] C. A. Kumins, *Ind. Eng. Chem. Anal.*, 1947, **19**, 376.
[2] F. Feigl, *Specific, Selective and Sensitive Reactions*, 1949, p. 214.
[3] R. B. Hahn, *Analyt. Chem.*, 1949, **21**, 1579.
[4] R. E. Oesper and J. J. Klingenberg, *ibid.*, 1509.

19. PHTHALIC ACID

Phthalic acid,

precipitates zirconium quantitatively from solutions up to 0·35 N in hydrochloric acid.[1] The acidity recommended for the precipitation is 0·3 N, since several other metals interfere below this value.

Above 0·35 N low results are obtained. At the specified acidity zirconium can be separated from aluminium, beryllium, cerium earth metals, iron, manganese, nickel, thorium and uranium by a single precipitation. Separation from chromium, tin, titanium and vanadium requires a second precipitation. Chromium and vanadium are not themselves precipitated even in neutral solution, but in their presence the precipitate of zirconium is slightly coloured, although the weight of the resulting zirconium oxide is not in excess of the expected value. After a second precipitation the oxide is pure white. Apart from iron, thorium, tin and titanium, none of the other elements precipitates in neutral solution, so that many valuable separations may therefore be effected. With increasing acidity the precipitate becomes more gelatinous and takes longer to settle, but flocculation is assisted by the presence of ammonium nitrate. The final determination is made by ignition to the oxide.

Procedure

Determination of zirconium. Add 30 ml. saturated ammonium nitrate solution to the chloride solution containing zirconium. Add sufficient 2 N hydrochloric acid to give a solution that will be 0·3 N in this acid when diluted to 200 ml. Dilute the solution to 100 ml. with water, and boil. Add, whilst stirring continuously, 100 ml. boiling 4 per cent aqueous phthalic acid solution, and boil gently for 2 minutes. Keep the solution on a boiling water-bath for 2 hours, and then set aside to cool for 1 hour.

Filter on a Whatman No. 42 filter paper, wash once with hot 0·1 per cent phthalic acid in 0·3 N hydrochloric acid, and then with 0·1 per cent phthalic acid in 2 per cent ammonium nitrate solution. Ignite and weigh as ZrO_2.

$$\text{mg. ppt.} \times 0·7403 \equiv \text{mg. Zr.}$$

Determination of zirconium in the presence of titanium or tin. Dissolve the precipitate, before ignition, in 1 : 1 hydrochloric acid, dilute, and neutralise with ammonia, using Congo Red as indicator. Add the calculated amount of 2 N hydrochloric acid, and reprecipitate in the same manner as before, completing the determination as previously described. A slight precipitate may appear during the neutralisation, but this will disappear when the acid is added.

[1] A. Purushottam and B. S. V. R. Rao, *Analyst*, 1950, **75**, 684.

20. SALICYLIMINE

When salicylaldehyde and certain of its derivatives are dissolved in aqueous ammonia or in aqueous methylamine solution, rather unstable derivatives, which are stated to be salicylimines, are formed:

The solutions, without isolation of the products, give coloured derivatives with a number of cations, and in particular with nickel and copper.[1] Salicylimine itself, prepared thus in solution, may be used for the gravimetric determination of copper or of nickel.

The composition of the complexes formed appears to correspond to the formula $M(Sal)_2$, where H.Sal is taken as representing salicylimine. The best results are obtained at a pH of 7·0–12·0, and the reagent is unstable in weakly acid solutions. The precipitate should not be heated above 105° C.

Iron-II, iron-III, palladium and vanadium also give precipitates, and so will interfere, but cadmium, manganese, mercury and zinc, in addition to a number of others of the commoner cations, do not precipitate.

Solution required

Reagent solution. Dissolve 1 g. redistilled salicylaldehyde (b.p. 195°–196° C.) in 100 ml. 1·5 N aqueous ammonia.

Procedure

Determination of copper in brass or bronze. Weigh accurately 60–100 mg. brass or bronze, and dissolve in 4 ml. 1 : 1 hydrochloric-nitric acid mixture. Dilute to approximately 25 ml. with water, and add 25 ml. 10 per cent sodium tartrate solution. Cool below 25° C., make distinctly ammoniacal with filtered ammonium hydroxide, cool again if necessary, and add excess of the reagent solution.

Set the cold solution aside for 5 minutes, filter, and wash the precipitate. Dry for 1 hour at 100° C.

$$\text{mg. ppt.} \times 0·2092 \equiv \text{mg. Cu.}$$
$$(\text{mg. ppt.} \times 0·1963 \equiv \text{mg. Ni.})$$

[1] F. R. Duke, *Ind. Eng. Chem. Anal.*, 1944, **16**, 750.

21. SODIUM NAPHTHIONATE

Thorium can be separated from associated cerium earth elements by a single precipitation with sodium naphthionate:[1]

The exact acidity at which the precipitation is carried out is important, because when this is too high precipitation is incomplete, and when it is too low the precipitate is flocculent and difficult to free from impurities by washing. Accordingly the pH should be maintained between 2·3 and 3·2. Amounts of thorium between 9 and 200 mg. may be determined satisfactorily.

Cerium earth metals may be determined in the filtrate by making the solution alkaline with ammonia, filtering, washing and igniting. The residue is weighed as R_2O_3. A correction is made for oxygen in excess of the stoicheiometric amount by dissolving in hydrochloric acid, adding potassium iodide, and titrating the liberated iodine.

Procedures

Determination of thorium. Make the solution, containing not more than 0·2 g. thoria, just acid to Congo red. Dilute to 100 ml., and boil. Add slowly, whilst stirring, 100 ml. boiling 10 per cent sodium naphthionate solution. A pink crystalline precipitate forms. Boil for a further 5–10 minutes, and set aside to cool. Filter through a Whatman No. 42 filter paper, wash the precipitate three times by decantation with cold water, and finally four or five times on the filter paper. Ignite wet, and weigh as ThO_2.

$$\text{mg. ppt.} \times 0\cdot8788 \equiv \text{mg. Th.}$$

Determination of thorium in monazite. Dissolve the mixture of thoria and lanthanon oxides, obtained from a 30 g. sample according to the procedure of Miner[2] and freed from phosphate by double precipitation with oxalic acid, in concentrated nitric acid. Add a few ml. hydrogen peroxide to convert cerium-IV to cerium-III.

Evaporate almost to dryness on a water-bath. Dissolve the residue in water, and make up to 500 ml.

Take 25 ml. of this solution and carry out the determination of thorium as already described.

[1] A. Purushottam and B. S. V. R. Rao, *Analyst*, 1950, **75**, 555.
[2] H. S. Miner, *U.S. Bur. Min., Bull. No.* 212, p. 53, 1923.

22. TANNIN

Normally germanium is separated from other elements by distillation as the tetrachloride, b.p. 86°C. Tin is prevented from distilling by addition of sulphuric acid to the solution being distilled, and arsenic may be retained by carrying out the distillation in a current of chlorine, thus converting the trichloride to the less volatile pentachloride.

An alternative method, which avoids the necessity for distilling in chlorine, makes use of the fact that germanium may be determined in the presence of moderate amounts of arsenic by precipitation with tannin.[1] This method has been applied to the determination of germanium in steels.[2] No interference from any element normally present in steels has been observed.

Solution required

Wash liquid. Dissolve 50 g. ammonium nitrate and 5 g. tannin in water. Add 5 ml. concentrated nitric acid and make up to 1 litre with water.

Procedure

Determination of germanium in steel. Use all-glass apparatus. Fit a 500-ml. round-bottomed flask, through a ground-glass joint, with an efficient fractionating column which leads, through a water condenser and a safety bulb (to avoid suck-back) to 100 ml. water in a beaker which stands in an ice-bath. At the top of the fractionating column fit a side arm with a short length of rubber tubing, closed by a spring clip, as an additional safeguard.

Weigh 1–10 g. sample (for 5·0–0·5 per cent germanium) into the round-bottomed flask. Add a mixture of 10 ml. concentrated nitric acid and 100 ml. 1 : 4 sulphuric acid to the sample, and when most of the reaction has ceased, boil gently for 30 minutes. Cool the solution and add 5 g. copper turnings to react with the excess nitric acid. Boil for 3 minutes to expel nitrogen oxides.

Wash down the flask with a little water, add 200 ml. concentrated hydrochloric acid, and immediately connect the flask to the fractionating column. Distil, adjusting the heating rate so that 1 drop of constant boiling acid distils every 5 seconds. Collect 20–30 ml. distillate in the ice-cold water in the receiver. The germanium tetrachloride will have distilled almost completely before any hydrochloric acid comes over.

Disconnect the apparatus, and wash the inside of the safety bulb with water. Add 2 g. hydroxylamine hydrochloride to reduce any oxidising substances. Add 30 ml. fresh 5 per cent tannin solution, with stirring. Add a few drops methyl red indicator solution, followed by ammonia until the solution is alkaline, and bring just back to the acid side with sulphuric acid. Add 0·5 ml. concentrated sulphuric acid in excess. Heat just to boiling, and set aside to cool.

Filter through a Whatman No. 40 filter paper, and wash with the wash liquid until the precipitate is free from chlorides. Ignite in a platinum crucible, gently at first and then at 600° C. for 1 hour. Allow to cool, add 5 drops concentrated sulphuric acid and 3 ml. concentrated nitric acid, evaporate to dryness, burn off carbon below 600° C., and finally ignite at 900°–1000° C. for 10 minutes. Cool and weigh as GeO_2. The germanium dioxide should be quite white in colour.

$$mg.\ ppt. \times 0·6941 \equiv mg.\ Ge.$$

[1] G. R. Davies and G. T. Morgan, *Analyst*, 1938, **63**, 388.
[2] A. Weissler, *Ind. Eng. Chem. Anal.*, 1944, **16**, 311.

23. TETRAPHENYLARSONIUM CHLORIDE

Chretien and Longi[1] examined fifteen gravimetric methods for the determination of thallium, and concluded that of these only the determination as thallous chromate was worth retaining as a method. However, the solubility of the precipitate, even in this determination, is sufficiently great to require very strictly standardised conditions for the determination.

Smith[2] has since found that excellent results may be obtained by precipitating thallium-III with tetraphenylarsonium chloride, $(C_6H_5)_4AsCl$, the precipitate having the composition $(C_6H_5)_4AsTlCl_4$. The filtrate gives a negative test for thallium when treated with ammonium sulphide and with 10 per cent potassium iodide.

The precipitate must be washed with hydrochloric acid, as otherwise hydrolysis occurs and low results are obtained. Cations which form insoluble chlorides interfere, as do also the anions bromide, fluoride, iodide, nitrate, perchlorate, periodate, permanganate, perrhenate and thiocyanate.

Procedure

Determination of thallium. If the thallium is present as thallium-I, oxidise to thallium-III by adding 2 ml. 30 per cent hydrogen peroxide in the presence of sodium hydroxide solution. Acidify with hydrochloric acid, and add a few ml. concentrated acid in excess. A white precipitate will form which may be a thallium-I–thallium-III complex, but this will redissolve on addition of a further ml. hydrogen peroxide.

Dilute the solution to render it 0·5–2·0 N in hydrochloric acid, and add an excess of reagent solution containing 6·7 g. in 100 ml. water. The volume at this stage may vary between 25 and 75 ml. Heat to boiling to coagulate the white precipitate, and set aside overnight. Filter on a sintered-glass filter crucible. Wash with 20–40 ml. N hydrochloric acid, and dry in an oven at 110°C. Weigh as $(C_6H_5)_4AsTlCl_4$.

$$\text{mg. ppt.} \times 0·2802 \equiv \text{mg. Tl.}$$

[1] A. Chretien and Y. Longi, *Bull. Soc. chim.*, 1944, **11**, 241.
[2] W. T. Smith, *Analyt. Chem.*, 1948, **20**, 937.

24. TRIPHENYLMETHYLARSONIUM IODIDE

Triphenylmethylarsonium iodide, $(C_6H_5)_3 . CH_3 . AsI$, has been recommended for the gravimetric determination of cadmium in the presence of zinc.[1] The gravimetric method may be used over the range of 1–100 μg. cadmium per ml., lower concentrations being determined nephelometrically. Apparently the acid concentration is not highly critical, since for the nephelometric method it is stated that it may lie between 0·001 and 0·01 N.

If zinc is present to an extent of 100 times the amount of cadmium or more, potassium iodide must be added to provide further iodide ion, since the zinc forms the complex ion ZnI_4^{-}, reducing the iodide concentration in the solution and preventing complete precipitation of the cadmium.

Antimony, arsenic, bismuth, copper, lead, mercury and silver interfere, by forming either insoluble iodides or similar complexes. These elements may, however, be eliminated by boiling the acid solution with iron wire and filtering before the determination.[2]

Procedure

Determination of cadmium in presence of zinc. Heat 50 ml. solution containing cadmium to boiling, and slowly add sufficient 0·5 per cent reagent in 0·5 per cent potassium iodide solution to provide three equivalents of the reagent. Stir well during the addition. If zinc is present to the extent of 100 times the cadmium, add 10 ml. 10 per cent potassium iodide solution. Set the whole aside for 1 hour to cool, and then cool in running water for a further half hour. Filter on an X4 sintered-glass filter crucible, wash with 0·5 per cent potassium iodide solution, and dry at 105°–110°C. Weigh as $[(C_6H_5)_3 . CH_3 . As]_2CdI_4$.

$$mg.\ ppt \times 0·0890 \equiv mg.\ Cd.$$

[1] F. P. Dwyer and N. A. Gibson, *Analyst*, 1950, **75**, 201.
[2] A. Pass and A. M. Ward, *ibid.*, 1933, **58**, 667.

CHAPTER V

INDICATORS

A VERY wide range of new indicators falling into the usual categories into which these are normally divided has been proposed in the past twenty years. It would require more space than is available here to deal with all of these fully. Consequently some selection has been applied, and such of the indicators as appear to be of less importance have been summarised in tabular form, with appropriate notes. Only those indicators which seem to have a reasonably wide appeal have been dealt with more fully.

A. ACID-BASE INDICATORS

1. SIMPLE INDICATORS

Tetrabromophenol blue

This compound, first proposed for use as an acid-base indicator by Harden and Drake,[1] has the pH range:

pH 3·0 Yellow.
pH 4·6 Blue.

It is stated to be more satisfactory than bromophenol blue, since its solutions are not dichroic.

Indicator solution. Dissolve 100 mg. tetrabromophenol blue in 1·01 ml. 0·1 N sodium hydroxide, and add a little distilled water. Make up to 100 ml. with distilled water.

Use 1–3 drops per 10 ml. solution.

Mercury iodocyanide

This compound decomposes, with the separation of red mercuric iodide, in the presence of traces of acid, its pH range being stated to be about that of methyl orange.[2] It is particularly recommended as an acid indicator for the titration of acidic organic compounds, and in particular of ring nitro-compounds.

Indicator solutions. Mix equal amounts of solutions (*a*) and (*b*). Solution (*a*): Dissolve 12·6 g. mercuric cyanide in 200 ml. water. Solution (*b*): Dissolve 6·3 g. iodine and 12·6 g. potassium iodide in 200 ml. water.

Thallium-III chloride

Thallium-III chloride and other thallium-III salts have been recommended as indicators in acid-base titrations.[3] The sharp end-point obtained with them agrees with that obtained with methyl orange, and the colour change is:

Acid   Colourless.
Alkaline   Yellow.

Thallium-III is claimed to possess a special advantage over organic indicators in that it can be used for the titration of sulphuric, nitric or hydrochloric acids in the presence of oxidising agents which would destroy the normal indicators. Such oxidising agents are chlorine, bromine, hypochlorite, hypobromite or nitrite.

Bases must be estimated by back titration.

Lead dithizonate

The insolubility of the red complex formed by lead and diphenyl-thiocarbazone in acid solution has been proposed[4] as a means of indicating the end-point in acid-base titrations. A small amount of lead salt is added to the solution being titrated, and a small amount of a solution of dithizone in an immiscible organic solvent is shaken with this during the titration. As soon as the acid has been neutralised the dithizone solution will change colour owing to the formation of the lead complex.

pH < 4·0 Green.
pH > 4·0 Red.

The titration may be carried out in dark-coloured solutions which prevent the use of more conventional indicators, if a suitable organic solvent which does not extract the colouring matter is chosen. Because the colour change occurs at pH 4·0, the indicator cannot be used for the titration of weak acids.

It has been pointed out[5] that owing to the lack of selectivity of dithizone, traces of other metals may interfere by forming more stable complexes.

10

Indicator solutions. Solution (*a*): Dilute lead acetate solution. Solution (*b*): Dissolve 50 mg. diphenylthiocarbazone in 1 litre chloroform, ethyl acetate, amyl alcohol, chloroform or benzene. Add a few drops solution (*a*) and a few ml. solution (*b*).

Mercury chromithiocyanate

Potassium chromithiocyanate and mercury cyanide solutions, when mixed, form a turbidity in acid solution, due to precipitation of mercury chromithiocyanate, $Hg_3[Cr(CNS)_6]_2$. This turbidity disappears on increasing the pH.[6] The change is sensitive, and enables titrations to be carried out in dark-coloured solutions. It is not affected by oxidising agents, *e.g.*, nitric acid or small amounts of free chlorine or bromine. Free chromic acid may be determined in dichromate solution. Phosphoric acid may be titrated as a monobasic acid.

Bromides, cyanides, iodides, thiocyanates and thiosulphates interfere.

pH $<4\cdot0$ Turbidity.

pH $>4\cdot0$ No turbidity.

The potassium chromithiocyanate must be kept dry since it decomposes slowly in aqueous or alcoholic solution. It must also be kept free from excess thiocyanate. It is therefore dissolved as required in a solution containing mercuric cyanide and ammonium nitrate. The latter salt is present to prevent decomposition of the chromithiocyanate ion through sudden rise of alkalinity during titration.

Indicator solutions. (*a*) Stock solution: Dissolve 6 g. pure mercuric cyanide and 1 g. ammonium nitrate in 100 ml. water.

(*b*) Temporary solution: Dissolve $0\cdot02$ g. potassium chromithiocyanate in 10 ml. stock solution. This solution keeps for about 3 days.

Use 1 ml. temporary solution per 25 ml. of solution to be titrated.

Synthesis of potassium chromithiocyanate.[7] Dissolve 30 g. potassium thiocyanate and 15 g. chromic chloride, $CrCl_3 . 6H_2O$, in the minimum amount of water. Evaporate on the steam-bath. Extract the residue with absolute ethanol, concentrate, and filter. Wash with ether. Redissolve in ethanol, and filter off precipitated potassium chloride. Evaporate the solution. Dark red crystals of potassium chromithiocyanate, $K_3[Cr(CNS)_6]_2$ separate out. These give a blue-violet solution in water. Yield 12–15 g.

Recrystallise twice from ethanol, and store dry.

Benzoyl auramine G

This indicator has been recommended[8] as being more satisfactory for Kjeldahl nitrogen determinations than either cochineal or methyl red, which are usually recommended. It gives a very sharp colour change.

pH 5·0 Dichroic: blue in thin layer, red-violet in thick layer.
pH 5·6 Pale yellow.

Since the indicator hydrolyses slowly, both in acid and in alkaline solution, it should only be added immediately before the titration, which should then be carried out rapidly. The change at the end-point is equally readily observed by daylight or by artificial (tungsten) lighting, although the actual colours appear somewhat different.

Indicator solution. Dissolve 0·25 g. benzoyl auramine G in 100 ml. methanol.

Use 3–7 drops.

Ethyl-*bis*-(2 : 4-dinitrophenyl) acetate

This indicator,

is suitable for titrations in which phenolphthalein is normally used, but the colour change is such that it may be employed in amber-coloured solutions, or by operators with defective colour vision.[9]

pH 7·5 Colourless.
pH 9·1 Deep blue.

The indicator may be used in non-aqueous solvents such as ethanol, ether or benzene.

Indicator solution. A saturated solution in 50 : 50 ethanol-acetone.
Use 5 drops for 100 ml. solution.

Diphenylcarbazide

Diphenylcarbazide has been recommended[10] as an indicator for acid-base titrations, in dark-coloured solutions. To the acid solution is added 0·5 ml. diethyl ether, a few drops 35 per cent aqueous formaldehyde solution and then the indicator solution, and titration is carried out, with continuous shaking, until the ether shows the first colour change.

Indicator solution. A saturated solution in methanol.
Use 4–5 drops.

Pyridinium glutaconaldehyde perchlorate

This compound,

$$[O{=}CH{-}CH{=}CH{-}CH{=}CH{-}N \cdot C_5H_5]^+ClO_4^-$$

is colourless, but changes, within one unit of the pH scale, to a yellow compound which quickly becomes red.[11] As an indicator it is practically unaffected by the presence of neutral salts, or by the presence of ethanol up to 60 per cent.

Miscellaneous   TABLE III

Name	pH Range	Indicator Solution	Notes
Patent Blue V[12]	0·8 Yellow 1·2–2·0 Yellowish-green to green 2·0 Bluish tinge 3·0 Pure blue	0·1 per cent aqueous	1
Rubrophen (Trimethoxydihydroxy-oxotriphenylmethane[13]	1 Cherry red 3 Yellow 6·5 Yellow 7·2 Violet		2
Cymyl orange[14]	~3 Pure yellow ~4 Pink		3
p-Dimethylamino-p-azobenzene-sulphonamide[15]	3 Red 4·5 Yellow		4
p-Ethoxychryso-ïdine[16]	3·5 Yellow 5·5 Red	0·2 per cent ethanolic. Use 1 drop per 50 ml. solution	

TABLE III—*continued*

Name	pH Range	Indicator Solution	Notes
4-Nitrocatechol[17]	3·92 Straw yellow 6·34 Lemon yellow 10·57 Red	1 per cent aqueous	5
Disodium-4 : 4'-*bis* (*p*-dimethylamino-azo)-2 : 2'-stilbene disulphonate[18]	4·0 Orange-red 5·4 Bluish-violet	0·1 per cent aqueous. Use 2 drops per 100 ml. solution	6
Disodium-4 : 4'-*bis* (*o*-tolyltriazeno) -2 : 2'-stilbene disulphonate[18]	4 Yellow 5 Muddy	0·5 per cent aqueous. Use 5 drops per 100 ml. solution	7
Mercurochrome[19]	4·5 Pink (red precipitate 4·6 Fluores-cence	1 per cent aqueous. Use 1 drop for each titration	8
5 : 8-Quinolinequin-one-8-hydroxy-quinolyl-5-imide (Indo-oxime)[20]	6·0 Red 8·0 Blue		9
1 : 3 : 8-Trihydroxy-6-methylanthra-quinone (Emodin)[21]	6·8 Yellow 7·6 Orange red	0·1 per cent ethanolic. Use 1–2 drops	10
1-(2-Hydroxy-1-naphthylazo)-2-naphthol-4-sul-phonic acid[22]	7·0 Red 8·3 Marine blue		
Sodium quinizarin-6-sulphonate[23]	7·1 Yellow 9·1 Pink 11·5 Blue-violet		11
Benzene-2-hydroxy-benzanthrone[24]	<7·4 Yellow >7·4 Purple	Saturated ethanolic. Use 6 drops	
Benzene-2-oxobenz-anthrone[24]			Acts similarly to previous indicator
3 : 4-*bis*-(4-Hydroxy-3-nitrophenyl)-hexane[25]	7·6 Colourless 8·2 Yellow		12

TABLE III—*continued*

Name	pH Range	Indicator Solution	Notes
Disodium 2-(4'-nitrophenylazo)-1-naphthol-4 : 8-disulphonate[26]	8 Pink 10 Light purple	0·1 per cent aqueous	
Disodium 2-(2'-methoxy-4'-nitro-phenylazo)-1-naphthol-4 : 8-disulphonate[26]	9 Pink 11 Light purple	0·1 per cent aqueous	
Disodium 2-(4'-nitrophenylazo)-1-naphthol-3 : 8-disulphonate[27]	11 Pink 13 Light purple	0·1 per cent aqueous	

Notes

1. The colours are reputed to be very stable, and it is stated that over its pH range it may be used successfully for the colorimetric determination of pH.

2. The compound is slightly soluble in water and dissolves readily in alkali to give a violet solution. The double pH change causes it to be recommended for the titration of weak and strong bases in the presence of each other.

3. This indicator is useful over the same range as methyl orange. Since the colours are more distinct than those shown by methyl orange, it is claimed to be superior. The material is prepared by converting aminocymene to the sulphate, dehydrating and sulphonating with fuming sulphuric acid, and diazotisation and coupling with dimethylaniline.

4. The particular advantage of this indicator is that it passes from the red to the yellow colour directly, no intermediate orange being observed.

5. The change from straw yellow to lemon yellow is sharp, and the further change to red is gradual. The indicator may be used for the titration of strong acids with strong or weak bases. The colour change is not affected by nitric acid, but is sensitive to carbon dioxide. Typical titrations for which it is recommended are nitric acid against sodium hydroxide, and hydrochloric acid against sodium hydroxide,

ammonium hydroxide or sodium carbonate. In the last of these a slight excess of acid is added, the solution is boiled to free it from carbon dioxide, and back titrated with sodium hydroxide.

6. This indicator has been proposed for titration in yellow solutions, where the alkaline colour shows as a deep green.

7. This indicator has been recommended for the titration of very dilute sodium carbonate solutions, the titration being carried out in a white porcelain basin to show up the slightest change in shade.

8. The titration is preferably carried out from the alkaline to the acid side, and results in the titration of potassium hydroxide, sodium carbonate and sodium bicarbonate agree with those obtained using methyl orange. The change is reversible.

9. This indicator is particularly useful in the titration of dilute (0·01 N) mineral acid against alkali hydroxide solutions.

10. Like litmus, this indicator may be used in the form of a solution, or as impregnated papers. The paper shows a colour change with one drop of 0·005 N hydrochloric acid or 0·05 N sodium hydroxide. The indicator can be used for the titration of weak acids or bases, and its sensitivity is little affected by the presence of neutral salts.

The paper may be prepared by saturating strips of filter paper in 0·1 per cent ethanolic solution and drying in absence of ammoniacal vapours. Alternatively strips of filter paper may be saturated with 0·05 N sodium hydroxide solution and dried, after which they are saturated with a 0·1 per cent solution of emodin in 0·05 N sodium hydroxide solution and again dried.

11. The results obtained using this indicator agree very closely with those obtained with phenolphthalein as indicator when sodium hydroxide is titrated against hydrochloric, sulphuric, nitric, acetic or oxalic acids. The colour change is sharper and more easily observed than the phenolphthalein one, and consequently the indicator is recommended for the titration of strong bases against either weak or strong acids. The nature of the colour change also makes the indicator suitable for the colorimetric determination of pH.

12. This indicator is recommended for the colorimetric determination of pH over its range, as the increase in yellow colour is progressive.

2. SCREENED INDICATORS

These consist of a simple acid-base indicator to which a dyestuff has been added in order to accentuate the colour change at the end-point. The improvement is often due to the fact that the dyestuff filters out a portion of those wavelengths not appreciably absorbed by either form of the indicator. Although the colour change at the end-point may be completely different from that of the unscreened indicator, the pH interval over which the change takes place remains unaffected.

Methyl yellow-methylene blue

Carmody[28] claims that methyl yellow screened by methylene blue is superior to the screened methyl orange-xylene cyanole FF recommended by other authors.[29] First recommended by Kolthoff and Rosenblum,[30] the proportions of the screening dye have, after an investigation of absorption curves, been modified by Carmody, and the colour change obtained, with the proportions now recommended, is as follows:

> pH 2·9 Yellow-green.
> pH 3·3 Pink-straw.
> pH 4·0 Pink.

The first appearance of pink is taken as the end-point. Since methyl yellow is rapidly destroyed in alkaline solution, it is recommended that in order to avoid any effect on the end-point the addition of indicator should not be made until the titration is more than half completed. Direct titration of sodium carbonate with 0·02–0·05 N acid gives a very satisfactory end-point.

Indicator solution. Dissolve 0·8 g. methyl yellow and 0·04 g. methylene blue in 1 litre ethanol.

Use 0·1 ml. per 100 ml. final volume of titration solution.

Methyl orange-indigo carmine

Smith and Croad[31] recommend this indicator for the titration of sodium carbonate with acid. The end-point is

Alkaline	Green
Transition	Grey
Acid	Violet,

the change taking place at pH 4·1.

Indicator solution. Dissolve 0·10 g. methyl orange and 0·25 g. indigo carmine in 100 ml. water. Store in a dark glass bottle.

Methyl red-alphazurin

First investigated by Johnson and Green,[32] the proportions of this mixture have been modified by Fleisher[33] to make it suitable for the determination of the alkalinity of boiler-feed water. For this purpose the most suitable end-point is at pH 4·89. Methyl orange, with a pH range of 3·1–4·4, and methyl red (pH 4·2–5·4) screened with xylene cyanole FF have both been used, but neither has proved completely satisfactory. With the latter, the end-point is observed at a pH very much lower than would be expected from the range of the unscreened indicator. By using an appropriate indicator solution, Fleisher was able to obtain an end-point in the range

> pH>4·8 Green-grey
> pH   4·8 Grey
> pH   4·6 Purple-grey.

Two drops of 0·05 N hydrochloric acid are sufficient to cause the change from green-grey to purple-grey. Large quantities of neutral salt increase the apparent pH of the equivalence point, but have no effect on the stoicheiometric end-point.

Since ethanolic solutions were found to have poor keeping qualities, it is more satisfactory to prepare the indicator in aqueous solution, using the sodium salt of methyl red.

Indicator solution. Dissolve 0·45 g. sodium salt of methyl red and 0·55 g. alphazurin in 1 litre distilled water.

Tropolein 000 No. 1-malachite green

Krishnamurty[34] has tested several indicators in respect of suitability for the titration of carbonate to bicarbonate with hydrochloric acid. He found that a screened indicator consisting of equal parts of Tropolein 000 No. 1 and malachite green gave the best end-point.

The pH range of malachite green is

> pH 0·1   Yellow
> pH 2·0   Blue-green.

3. MIXED INDICATORS

Several advantages may be derived from the use of mixed indicator solutions. In particular, by choosing indicators showing complementary colours on the acid and alkaline side, or by varying the relative proportions of the indicators, the end-point may be made more distinct than any obtainable with a single indicator.

Bromocresol green-methyl red

Ma and Zuazaga[35] recommend this mixed indicator solution for titration in the determination of nitrogen by the Kjeldahl method, using boric acid as absorbent for the ammonia. The indicator shows the following colour changes:

Boric acid solution	Bluish-purple.
Ammonia	Bluish-green.
Acid	Pink.

Since the indicator has a different colour in boric acid solution from those shown in presence of traces of ammonia or of mineral acid, any contamination in the titration flask is immediately obvious on adding the indicator. In addition, if no change is observed when the first distillate reaches the boric acid it is clear that either the sample is nitrogen-free, or insufficient sodium hydroxide has been added to liberate the ammonia. The end-point is taken as the disappearance of the blue colour (or the appearance of a faint pink, with a correction of 0·02 ml.).

Indicator solution. Dissolve 0·1 g. bromocresol green in 100 ml. 95 per cent ethanol and 0·1 g. methyl red in 100 ml. 95 per cent ethanol. Mix 10 ml. bromocresol green solution with 2 ml. methyl red solution.

Use 4 drops per 5 ml. 2 per cent boric acid.

Bromocresol green-methyl yellow

As an indicator for use in the methyl orange range, particularly when coloured solutions are being titrated, Hoppner[36] recommends this mixed indicator. Tests showed that the indicator was very useful in double titrations where phenolphthalein and methyl orange are normally used. The change is sharp, and is as follows:

Alkaline	Blue.
Acid	Yellow.

The change from a dark to a light shade makes the end-point easy to observe.

Indicator solution. Dissolve 0·2 g. bromocresol green in 100 ml. ethanol, and 0·2 g. dimethyl yellow in 100 ml. ethanol. Mix 8 ml. bromocresol solution with 2 ml. dimethyl yellow solution.

Use 8 drops per 100 ml. solution.

Miscellaneous mixed indicators

Hähnel[37] has investigated sixteen binary combinations of indicators, or of indicators and dyes, and has found those listed below to be useful:

(a) 5 parts dimethyl yellow   + 3 parts methylene blue.
(b) 1 part methyl red   + 1 part bromocresol green.
(c) 5 parts neutral red   + 2 parts methylene blue.
(d) 5 parts neutral red   + 2 parts tetrabromophenol blue.
(e) 3 parts phenol red   + 2 parts bromothymol benzein.
(f) 3 parts cresol red   + 1 part thymol benzein.
(g) 3 parts phenol red   + 4 parts bromothymol blue.
(h) 1 part cresol red   + 4 parts thymol blue.

The first six of these are particularly recommended, while (g) and (h) are proposed as slightly less satisfactory substitutes for (e) and (f) respectively.

Indicator (a) is proposed for the titration of strong and weak bases or of 30 per cent ethanolic hydrazine. For the titration of alkali either with ammonium or with acids, (b) is suitable. For general titrations in aqueous solution (c) or (d) may be used, and (d) can also be used in the titration of ethanolic alkali with acids in solutions containing up to 70 per cent ethanol. For determining either the acidity or the ester content of up to 70 per cent ethanol, (e) or its substitute (g) may be used, and (f) or (h) are suitable for the titration of ethanolic alkali with biphthalate.

4. UNIVERSAL INDICATORS

1. This indicator covers the range from pH 1·2 to pH 12·7.[38] Dissolve in 1 litre anhydrous ethanol, or preferably methanol,

> 0·125 g.   s-trinitrobenzene
> 0·0355 g. phenolphthalein
> 0·03 g.   cresolphthalein
> 0·1 g.   bromothymol blue (dibromothymol-
>   sulphonephthalein)
> 0·022 g.   methyl red
> 0·0085 g. methyl orange
> 0·05 g.   pentamethoxyl red.

2. This indicator, for the low pH range, covers the region pH 1·0–7·0, and changes colour in the order of the spectrum, from red to blue, each unit of the pH range corresponding to one spectrum colour.[39] Dissolve in 1 litre 50 per cent ethanol

> 0·35 g. thymolsulphonephthalein
> 0·2 g.   Tropolein 00
> 0·1 g.   tetrabromophenolsulphonephthalein
> 0·3 g.   bromocresol green
> 0·4 g.   bromocresol purple.

3. An indicator for the high pH range (pH 7·0–14·0) can be prepared which shows colour changes following the same pattern as indicator No. 2.[39] Dissolve in 1 litre 50 per cent ethanol

> 0·35 g. neutral red
> 0·15 g. thymolsulphonephthalein
> 0·25 g. thymolphthalein
> 0·1 g.   nitramine
> 0·6 g.   m-nitrophenol.

4. An indicator covering the range pH 4·0–10·0 gives the spectral colour changes red, orange, yellow-green, blue, indigo and violet in steps of one pH unit, intermediate colour changes being recognisable with an accuracy of 0·5 pH unit.[40] Dissolve in 100 ml. 75 per cent ethanol

> 0·005 g. thymol blue
> 0·025 g. methyl red
> 0·060 g. bromothymol blue
> 0·060 g. phenolphthalein.

Neutralise with 0·01 N sodium hydroxide solution to a green colour, or for the high or low pH ranges use directly as prepared.

[1] W. C. Harden and N. L. Drake, *J. Amer. Chem. Soc.*, 1929, **51**, 562.
[2] Langhans, *Nitrocellulose*, 1939, **10**, 163, 184, 207: *C.A.*, 1940, **34**, 958.
[3] V. K. Zolotukhin, *Zavod. Lab.*, 1940, **9**, 133: *C.A.*, 1940, **34**, 5777.
[4] W. Hirsch, *Analyst*, 1948, **73**, 160.
[5] I. M. Kolthoff, *Analyt. Chem.*, 1949, **21**, 101.
[6] R. Uzel, *Coll. Czech. Chem. Comm.*, 1933, **5**, 457: *C.A.*, 1934, **28**, 1950.
[7] C. Mahr, *Z. anorg. Chem.*, 1932, **208**, 313: *C.A.*, 1933, **27**, 40.
[8] J. T. Scanlan and J. D. Reid, *Ind. Eng. Chem. Anal.*, 1944, **16**, 53.
[9] E. A. Fehnel and E. D. Amstutz, *ibid.*, 53.
[10] E. Kröller, *Deutsch. Lebensm. Rundschau*, 1948, **44**, 31: *C.A.*, 1948, **42**, 6264.
[11] G. Schwarzenbach, *Helv. Chim. Acta*, 1943, **26**, 418.
[12] J. H. Yoe and G. R. Boyd, *Ind. Eng. Chem. Anal.*, 1939, **11**, 492.
[13] L. Szebelledy and M. Ajtai, *Maygar Gyógyszeresztud Társaság Értesitöje*, 1937, **13**, 822: *C.A.*, 1938, **32**, 1608.
[14] A. S. Wheeler and J. H. Waterman, *J. Elisha Mitchell Sci. Soc.*, 1933, **49**, 36: *C.A.*, 1933, **27**, 5673.
[15] K. van Lente and G. Pope, *Trans. Illinois State Acad. Sci.*, 1948, **39**, 77.
[16] E. Schulek and P. Rózsa, *Z. anal. Chem.*, 1939, **115**, 185.
[17] S. R. Cooper and V. J. Tulane, *Ind. Eng. Chem. Anal.*, 1936, **8**, 210.
[18] M. Taras, *Analyt. Chem.*, 1947, **19**, 339.
[19] J. W. Airan, *Nature*, 1947, **160**, 88.
[20] R. Berg and E. Becker, *Z. anal. Chem.*, 1940, **119**, 81.
[21] Z. M. Umanskii, *Zavod. Lab.*, 1945, **11**, 404: *C.A.*, 1946, **40**, 1412.
[22] H. Singu and H. Masuo, *J. Chem. Soc. Japan*, 1939, **60**, 595.
[23] J. H. Green, *Ind. Eng. Chem. Anal.*, 1942, **14**, 249.
[24] F. H. Fish and W. H. Wrenn, *Virginia J. Sci.*, 1942, **3**, No. 1, 12: *C.A.*, 1942, **36**, 3114.
[25] L. Spitzer, *Ann. Chim. appl.*, 1942, **32**, 180: *C.A.*, 1943, **37**, 1343.
[26] K. H. Ferber, *Ind. Eng. Chem. Anal.*, 1946, **18**, 631.
[27] S. R. Cooper and V. J. Tulane, *ibid.*, 1936, **8**, 210.
[28] W. R. Carmody, *ibid.*, 1945, **17**, 141.
[29] G. F. Smith and G. F. Croad, *ibid.*, 1937, **9**, 141.
[30] I. M. Kolthoff and C. Rosenblum, *Biochem. Z.*, 1927, **189**, 26.
[31] G. F. Smith and G. F. Croad, *Ind. Eng. Chem. Anal.*, 1937, **9**, 141.
[32] A. H. Johnson and J. R. Green, *ibid.*, 1930, **2**, 2.
[33] H. Fleisher, *ibid.*, 1943, **15**, 742.
[34] K. V. S. Krishnamurty, *J. Indian Chem. Soc.*, 1948, **25**, 492.
[35] T. S. Ma and G. Zuazaga, *Ind. Eng. Chem. Anal.*, 1942, **14**, 280.
[36] K. Hoppner, *Deutsch. Zuckering.*, 1936, **61**, 361: *C.A.*, 1936, **30**, 5523.
[37] S. Hähnel, *Svensk. Kem. Tidskr.*, 1935, **47**, 4: *C.A.*, 1935, **29**, 2470.
[38] F. Čuta and K. Kámen, *Chem. Listy*, 1936, **30**, 22, 192: *Coll. Czech. Chem. Comm.*, 1936, **8**, 395: *C.A.*, 1936, **30**, 8064.
[39] J. V. Dubsky and A. Langer, *Chem. Obzor.*, 1936, **11**, 29: *C.A.*,
[40] J. H. N. van der Burg, *Chem. Weekblad*, 1939, **36**, 101.

B. REDOX INDICATORS

The literature on oxidation-reduction indicators up to 1938 has been extensively reviewed,[1] while Kolthoff[2] has covered briefly the main developments in the field since 1948.

(i) REDOX INDICATORS FOR GENERAL PURPOSES

1 COMPLEXES OF 1 : 10-PHENANTHROLINE AND ITS DERIVATIVES

1 : 10-Phenanthroline-ferrous complex

The compound 1 : 10-phenanthroline

forms a complex with the ferrous ion which has a very strong red colour, and which can be reversibly oxidised to a pale blue compound.[3]

$$(C_{12}H_8N_2)_3Fe^{++} \rightleftharpoons (C_{12}H_8N_2)_3Fe^{+++} + e^-.$$

It is a sensitive indicator for oxidation-reduction titrations, and is particularly valuable in titrations involving the ceric ion.

The transition E.M.F. (with reference to the hydrogen electrode as zero) is 1·06 volt, but if the colour change is observed visually the end-point is effectively 0·06 volt higher than this, since approximately 90 per cent of the indicator must be oxidised in order to repress the red colour.

Indicator solution. Dissolve 5·9465 g. 1 : 10-phenanthroline and 2·7802 g. ferrous sulphate, $FeSO_4.7H_2O$ in 1 litre distilled water.

Use 1 small drop of this solution for each titration.

1 : 10-Phenanthroline derivatives

The ferrous complexes of 5-nitro-1 : 10-phenanthroline and 5-methyl-1 : 10-phenanthroline have been found by Smith and Richter[4] to be suitable as reversible indicators. The colour change in each case is like that with the 1 : 10-phenanthroline complex. The transition E.M.F.s in molar acid are 1·25 and 1·02 volt respectively.

The poly-substituted derivatives cover a wide range of oxidation potentials.[5] Two of these, 4 : 7-dimethyl-1 : 10-phenanthroline and 3 : 4 : 7 : 8-tetramethyl-1 : 10-phenanthroline, have been studied in detail in relation to the ferrous-ferric and chromic-dichromate systems.[6] The former indicator (oxidation potential in 0·5 M acid= 0·88 volt) is recommended for titration of ferrous-dichromate in 0·5 M sulphuric or hydrochloric acid, and the latter (oxidation potential in 0·1 M acid=0·85 volt) for the same titration in 0·1 M acid.

Indicator solutions. Prepare as in the case of the ferrous-1 : 10-phenanthroline complex, but use the weight of 1 : 10-phenanthroline derivative shown in Table IV.

TABLE IV

Derivative	Weight required for 1 litre 0·01 M Fe++, g.
5-nitro-1 : 10-phenanthroline	6·7560
5-methyl-1 : 10-phenanthroline	6·3673
4 : 7-dimethyl-1 : 10-phenanthroline	6·7877
3 : 4 : 7 : 8-tetramethyl-1 : 10-phenanthroline	7·6290

2 : 2′-Dipyridyl-ferrous complex

This has been used[7] as a sensitive indicator for the reversible titration of ferric solutions after reduction in a Jones reductor. The colour change is very sharp, from a marked orange to practically colourless. Titration with an oxidising agent after reduction in a silver reductor, using N hydrochloric acid solution, produces a change from orange to yellow, the residual colour being due to complex ferric chloride.

Either the perchlorate or, less satisfactorily, the sulphate, may be used. In the latter case the titration is less accurate and the blank less reproducible.

Indicator solution. (a) *Perchlorate.* Dissolve 1·65 g. 2 : 2′-dipyridyl-ferrous perchlorate in 1 litre water. Use 1 ml. of this solution for each 200–300 ml. titration solution.

(b) *Sulphate.* Dissolve 6·9505 g. pure ferrous sulphate, $FeSO_4.7H_2O$, and 11·7135 g. 2 : 2′-dipyridyl in 1000 ml. water. Use 1 drop for each titration.

Synthesis of ferrous perchlorate complex. Add 11·7135 g. 2 : 2′-dipyridyl to a freshly prepared solution of 6·9505 g. pure ferrous sulphate, $FeSO_4 . 7H_2O$. Dilute to 300 ml. in a 400-ml. beaker, and add 72 per cent (1 : 9) perchloric acid until precipitation of the complex salt is complete. Filter off the red insoluble precipitate on a sintered-glass filter, wash well with water, and dry at 90°–100° C.

2 : 2′-Dipyridyl-ruthenium complex

When this complex, in the form of the nitrate, which is yellow in colour, is oxidised, the resulting form is colourless.[8] The potential of the indicator is close to that of the ceric ion in sulphate solution (1·44 volt in M sulphuric acid), so that the indicator is not satisfactory in ceric sulphate titrations. Ceric nitrate in nitric acid solution has a higher oxidation potential (1·65 volt in 1 M nitric acid) and in titrations in nitric acid solution sharp and reversible end-points are obtained.

The indicator is particularly recommended for the direct titration of sodium oxalate (in 2 M perchloric acid) with 0·1 M ceric nitrate.

Indicator solution. Prepare a 0·02 M solution of 2 : 2′-dipyridyl-ruthenium nitrate. Use 2 drops per 100 ml. solution.

The 1 : 10-phenanthroline-ruthenium complex has also been proposed[9] as a reversible redox indicator.

2. DIPHENYLAMINE DERIVATIVES

The barium salt of diphenylamine sulphonic acid has been shown[10] to be more satisfactory for dichromate titrations than diphenylamine sulphonic acid itself. The indicator properties of various diphenylamine derivatives have been investigated.[11] In each case the oxidised form of the indicator is red, and the reduced form is colourless. The colour changes are reversible and reasonably stable, although the more unstable oxidised form ought not to be allowed to stand for too long a period.

2-CARBOXY-2′-METHOXY DIPHENYLAMINE is especially recommended for the ferricyanide-vanadyl system. Direct titration of solutions of thiocyanate, thiosulphate and trivalent antimony with hypobromite in alkaline solution may be carried out using DIPHENYLAMINE-4-SULPHONIC ACID (first oxidised in acid solution) or 2-AMINO DIPHENYLAMINE-4-SULPHONIC ACID. Either of the first two of these three diphenylamine derivatives, or alternatively, either of the derivatives 2-CARBOXY-2′-METHYL DIPHENYLAMINE, 2-CARBOXY

DIPHENYLAMINE or 2 : 2′-DICARBOXY DIPHENYLAMINE are recommended for use in alkaline solution.

Kirsanov and Cherkassov[12] found that introduction of the carboxyl radicle into diphenylamine increased the oxidation potential of the indicator. The 2 : 2′-, 2 : 3′- and 2 : 4′-DICARBOXYLIC ACIDS function satisfactorily as indicators in strongly acid (16–20 N sulphuric acid) solution.

N-METHYL DIPHENYLAMINE-4-SULPHONIC ACID, which is water soluble, is claimed[13] to be more stable towards oxidising agents than the unsubstituted sulphonic acid. The colour change is from colourless to purple-red in dilute acid solutions. The reagent may be prepared by heating N-METHYL DIPHENYLAMINE with concentrated sulphuric acid at 150°C.

The indicator is suitable for use in the titration of ferrous iron with dichromate, permanganate or sulphatocerate. In N sulphuric acid or hydrochloric acid solutions the colour change occurs around 0·5–0·6 volt (with reference to the saturated calomel electrode).

Cohen and Oesper[14] have reported that a DIPHENYLAMINE MONO-SULPHONATE very suitable for ferrous-dichromate titrations can be prepared by treating diphenylamine either with very old samples of ethyl sulphate, or preferably with *n*-butyl sulphate. A deep purple colour, stable for at least 15 minutes, is obtained with slight excess of dichromate, and this is reversible, being discharged sharply by slight excess of ferrous ion.

Indicator solution. 0·1 per cent aqueous diphenylamine monosulphonate.

Use 4–5 drops for 400 ml., together with 15–25 ml. of the usual sulphuric-phosphoric acid retarding solution.

Synthesis of diphenylamine monosulphonate. Heat 21 g. (0·1 mole) *n*-dibutyl sulphate and 16·9 g. (0·1 mole) diphenylamine together on a boiling water-bath for 1 hour, and then on a metal-bath at 130°–140°C. for 1 hour. Add a solution of 4·8 g. sodium in 200 ml. ethanol, and reflux till the dark tarry reaction mixture becomes white. Distil off the ethanol. Take up the residue in water, and filter. Extract the filtrate thoroughly with ether to remove residual diphenylamine. Warm the aqueous solution and bubble air through it to remove ether. Add sodium hydroxide solution cautiously till a faint turbidity appears. Set aside for 24 hours. The sodium salt precipitates slowly. Filter this off, and suck as dry as possible. Dissolve in warm ethanol, and reprecipitate by the addition of excess of ether. Filter off, dissolve in warm ethanol, bubble in carbon dioxide until no more sodium carbonate separates, and concentrate on the water-bath till crystallisation commences.

3. BENZIDINE AND RELATED COMPOUNDS

Benzidine acetate

Benzidine acetate has been proposed[15] as an internal indicator for the determination of ferrocyanide with potassium dichromate. The colour change is from yellow-green to deep green, with precipitation. Since the acidity must be strictly controlled, it is found preferable to prepare standard potassium dichromate solution containing a calculated amount of acid, and the most satisfactory solution is one which is 0·02 N in potassium dichromate and contains 150 ml. concentrated hydrochloric acid per litre.

Indicator solution. 0·4 per cent aqueous.

Use 1 ml. for each 25 ml. ferrocyanide solution.

Naphthidine (4 : 4'-Diamino-1 : 1-dinaphthyl)

The naphthalene analogue of benzidine was first employed[16] as a reversible redox indicator in the titration of ferrous iron with dichromate. The oxidised form is deep red, the colour being discharged immediately by excess of ferrous ion. The coloured form is stable for one and a half to two minutes after the first drop of oxidising agent in excess has been added, fading gradually.

Mercury salts do not interfere with the behaviour of the indicator, and this allows it to be used for the titration of ferrous iron which has been obtained from the ferric form by reduction with stannous chloride.

The same indicator, in acetic acid medium, is suitable as an indicator in the titration of zinc with potassium ferrocyanide.[17]

Indicator solution. 1 per cent in concentrated sulphuric acid.

Use 3 drops for a total of 250 ml.

3 : 3'-Dimethylnaphthidine

This has been found[18] to be even more sensitive for the titration of zinc by potassium ferrocyanide than naphthidine, and to give a more stable oxidised colour, lasting for several days, the colour change being from greyish green to purple red.

It has proved more satisfactory in the titration of 0·001 M solutions than naphthidine in titrating 0·01 M solutions. For stronger solutions the acid (sulphuric acid) concentration should be N, and the solution should be adjusted, using 10 per cent ammonium sulphate solution, to give a concentration of 2 per cent of this salt. In more dilute solutions (less than 0·01 M) it is necessary to lower the concentration of these two constituents proportionately.

Indicator solution. Dissolve 0·2 g., with ,warming, in 100 ml. glacial acetic acid.

Add 4 drops 1 per cent potassium ferricyanide solution (freshly prepared each day and stored in a dark bottle) and 2 drops indicator solution for 50 ml. initial volume of 0·05 M solution. For more dilute solutions reduce the amount (or the strength) of the potassium ferricyanide solution to maintain the same relative proportions.

The same indicator, or, less satisfactorily, naphthidine or *o*-dianisidine, may be used[19] for the titration of cadmium, calcium or indium with ferrocyanide. For cadmium, the titration is carried out in neutral solution in the presence of ammonium sulphate (10 ml. 10 per cent ammonium sulphate solution added to 1–10 ml. cadmium solution, and the whole diluted to 40 ml.). Either direct titration or addition of excess ferrocyanide followed by back titration with standard cadmium solution may be employed. In the forward titration the colour change is from red-violet to white.

Calcium is titrated in the presence of ethanol (the initial concentration being 85–90 per cent and the final concentration 80–85 per cent). The colour change is from pink to pale yellow. An acidity not greater than 0·05 N hydrochloric acid can be tolerated.

Indium is titrated in neutral solution without ammonium sulphate present, as this appears to alter the composition of the ferrocyanide precipitated. Whilst the titration may be carried out directly, it is preferable in this case to add excess and back titrate. The colour change is from grey-white to pink-violet.

Indicator solution. Dissolve 0·2 g., with warming, in 100 ml. glacial acetic acid.

For cadmium use 2 drops 1 per cent potassium ferricyanide solution and 2 drops indicator solution for 40 ml. initial volume.

For calcium use 1–2 ml. 1 per cent potassium ferricyanide solution and 2 drops indicator solution.

For indium use 4 drops potassium ferricyanide solution and 2 drops indicator solution.

Naphthidine sulphonic acid

This indicator may be used in a variety of titrations[20] such as the titration of zinc with ferrocyanide. However, for this and similar titrations dimethylnaphthidine sulphonic acid is superior.

3 : 3′-Dimethylnaphthidine sulphonic acid

This indicator is somewhat better than naphthidine sulphonic acid for the titration of zinc with ferrocyanide, and is superior to 3 : 3′-dimethylnaphthidine itself down to 0·001 M zinc.[20] In the direct titration the colour change is instantaneous, being from green to deep reddish-purple.

In cadmium titrations the colour change is rather more intense than with naphthidine sulphonic acid, and is sharp. In ferrous-dichromate titrations the indicator has no particular advantage, but for the ferrous-vanadate titration a much more stable colour is obtained, so that the indicator can be added at the beginning of the titration.

Indicator solution. Dissolve 1 g. acid in a slight excess of aqueous ammonia, and boil to expel excess ammonia. Dilute to 100 ml. with water.

In titrating zinc use 4 drops 1 per cent potassium ferricyanide solution and 2 drops indicator solution. For cadmium use 2 drops ferricyanide solution and 2 drops indicator solution. For iron-II titrations use 4 drops indicator solution.

(ii) REDOX INDICATORS FOR SPECIAL PURPOSES

1. SODIUM HYPOCHLORITE TITRATIONS

Bordeaux B (British Colour Index No. 88)

This indicator has been recommended by Kolthoff and Stenger[21] as an indicator for the titration of reductants requiring an alkaline medium (*e.g.*, arsenic-III, ammonium salts, urea, sulphides, thiosulphates, tetrathionates, hydrogen peroxide and nitrates) with standard hypochlorite solution (p. 180). The indicator is destroyed at the end-point, so that it is irreversible. The colour change is from pink to colourless or to a faint yellow-green.

An indicator blank equivalent to 0·03 ml. 0·1 N hypochlorite solution should be deducted for each 0·1 ml. indicator used in a final volume of 50–75 ml.

Indicator solution. 0·2 per cent aqueous.

Amaranth (B.C.I. No. 184)

This dye has been found by Belcher[22] to be equally satisfactory to Bordeaux B as an irreversible indicator in hypochlorite titrations. The colour change is from red to colourless.

Tartrazine

Tartrazine has been proposed[22] as a reversible redox indicator for the titration of reductants requiring an alkaline medium (*e.g.*, arsenic-III, ammonium salts, urea, sulphides, thiosulphates, tetra-thionates, hydrogen peroxide and nitrates) with standard hypochlorite (p. 180). The indicators normally used for such titrations, such as Bordeaux B, have the disadvantage of being completely destroyed at the end-point. Some destruction of the tartrazine takes place, but if not more than 0·3 ml. 0·05 hypochlorite solution is added after the end-point is reached, the end-point can be passed and repassed several times. It is preferable to delay adding the indicator until near the end of the titration. The colour change is from yellow to colourless.

Indicator solution. 0·05 per cent aqueous.

Use 10 drops added towards the end of the titration for an initial volume of 25 ml., or add 5 drops at the beginning of the titration, titrate rapidly to faint yellow, and add a further 5 drops before completing the titration.

An indicator blank of 0·05 ml. 0·05 N sodium hypochlorite solution should be deducted for 10 drops indicator solution when the final volume is 30–60 ml.

Quinoline yellow

Under suitable conditions quinoline yellow may be used[23] as a reversible indicator in hypochlorite-arsenite titrations (p. 180).

To the arsenite to be titrated 0·5 g. solid potassium bromide is added followed by the indicator. The solution is then titrated with standard (0·05 N) sodium hypochlorite solution. The colour change, which is from yellow to colourless, is reversed on adding standard arsenite solution.

Indicator solution. 0·2 per cent aqueous.

Use 10 drops indicator solution.

2. POTASSIUM IODATE TITRATIONS

Amaranth (B.C.I. No. 184)

Brilliant Ponceau 5 R (B.C.I. No. 185)

Naphthol Blue Black (B.C.I. No. 246)

The colour of these three dyes is bleached by a trace of excess iodate, and they have been used as indicators by Smith and Wilcox[24] in the Andrews iodate-iodine monochloride titration, in place of the immiscible solvent more usually added to the solution to indicate the end-point by disappearance of the iodine colour (excess of potassium iodate in the presence of a large amount of hydrochloric acid oxidising the iodine to iodine monochloride). Since the immiscible solvent is avoided, stoppering and shaking the titration flask is rendered unnecessary, and the titration is carried out in the normal manner.

Indicator solution. 0·2 per cent aqueous.

p-Ethoxychrysoïdine

This indicator functions reversibly in the titration of arsenite by iodate according to the procedure of Smith and Wilcox. The reversibility is limited, owing to partial destruction, so that at most the titration can only be reversed three times in succession.[25] Just before the true end-point the red colour of the indicator changes to a deep purple, then turning, on addition of further titrant, to orange at the end-point. In the second titration the end-point is equally sharp, but a third titration gives a less distinct change.

Indicator solution. 0·1 per cent ethanolic.

Add 12 drops indicator solution when fading of the colour shows that the end-point is being approached.

3. POTASSIUM BROMATE TITRATIONS

Chrysoïdine (B.C.I. No. 21)

Bordeaux B (B.C.I. No. 88)

Naphthol Blue Black (B.C.I. No. 246)

For many years the presence of free bromine at the end-point of bromometric titrations has been detected by the bleaching of methyl orange, methyl red or indigo carmine, the change not being reversible, since the indicators are destroyed.

The three indicators Chrysoïdine, Bordeaux B and Naphthol Blue Black, have been recommended instead of the more orthodox indicators.[26]

Indicator solutions. Dissolve 0·1 g. of the dyestuff in 100 ml. distilled water.

Use a few drops for 100 ml. solution. In order to avoid mistaking the end-point a further drop or two is added at or near the end of the reaction.

Brilliant Ponceau 5 R (B.C.I. No. 185)

This has been recommended,[26] in addition to Bordeaux B and Naphthol Blue Black, for the titration of trivalent arsenic or antimony in hydrochloric acid solution by bromate at room temperature. The intense colour of the dyestuffs permits titration of dilute solutions. The potentials at which they are oxidised with the destruction of the colour, using hydrochloric acid solutions of bromate as oxidant, are above the equivalent point potentials for the oxidation of the trivalent elements to the pentavalent states.

The indicators are irreversible, being destroyed by the oxidant.

Indicator solutions. 0·2 per cent aqueous.

Use a few drops per 100 ml. solution, and, as before, add one or two further drops near the end-point.

Fluorescein

This functions as an irreversible indicator in titrating arsenite with bromate.[27] At the end-point the greenish-yellow colour of the solution changes to a reddish-brown. The transition is slow in the cold, but heating to 40°–50° C. accelerates the reaction.

Indicator solution. Dissolve 0·1 g. fluorescein in 100 ml. distilled water containing a few drops of sodium hydroxide solution.

Use 1 drop for 10 ml. solution.

1-Naphthoflavone

Uzel[28] first showed that 1-naphthoflavone could be used as a reversible indicator for bromate titrations. The substance has

recently been examined critically by Belcher,[29] who finds that for
the titration of arsenite with bromate it is preferable to others
examined. On addition to the indicator a colloidal solution is formed
which is pale yellow. Excess bromine changes this colour to deep
orange brown, the colour change taking place on the surface of the
colloid. Addition of excess arsenite restores the original pale yellow
colour.

The indicator may be applied to the indirect determination of
small amounts of aluminium through 8-hydroxyquinoline. The metal
is precipitated as the 8-hydroxyquinolate, and the washed precipitate
is dissolved in hydrochloric acid. Potassium bromide solution and
an excess of standard bromate solution are then added. The excess of
bromate is determined by adding an excess of standard arsenite
solution and back titrating with standard bromate. Direct bromination of the 8-hydroxyquinoline solution is too slow to give a satisfactory end-point. With amounts of aluminium greater than 5 mg.
the colour of 8-hydroxyquinoline obscures the indicator change.

Indicator solution. 0·2 per cent in ethanol.

Use 0·3–0·5 ml. For the determination of aluminium use 1 ml.,
or more if the solution is highly coloured.

p-Ethoxychrysoïdine

This compound has been recommended as an indicator in acidi-
metry, argentometry and bromometry,[30] and, as already stated
(p. 150), may be used in iodate titrations. It has also been used
for the titration of zinc with potassium ferrocyanide.

It may be used reversibly in the titration of bromate with arsenite.[29]
A solution of potassium bromide in sulphuric acid becomes red on
addition of the indicator. The red colour deepens on addition of a
drop of dilute potassium bromate solution, but further addition of
the oxidising agent bleaches the colour to orange-yellow. Addition
of excess arsenite solution reconverts the indicator to the red form
although there appears to be some destruction of indicator, the first
end-point being the sharpest.

p-Ethoxychrysoïdine is also suitable for use in titrations of
trivalent antimony with bromate, being preferable for this purpose
to 1-naphthoflavone, which appears to be affected by the presence
of the required tartaric acid.

Indicator solution. 0·05–0·1 per cent solution in ethanol.

Use 2–3 drops.

Quinoline yellow

This indicator, which has been recommended for hypochlorite-arsenite titrations (pp. 149, 180), may be used[31] for the titration of arsenic or antimony bromometrically with standard potassium bromate solution. The indicator is reversible, but is slowly destroyed, so that it is advisable to delay addition until near the end of the titration.

Indicator solution. 0·5 per cent aqueous.

Use 8 drops indicator solution. Preferably add 4 drops at the beginning of the titration and a further 4 drops when the colour begins to fade.

(iii) MISCELLANEOUS REDOX INDICATORS

Toluylene Blue

Toluylene blue in dilute sulphuric acid solution has been used as a reversible indicator for the titration of ferrous ion with cerate.[32] Mercuric, stannous and stannic ions and hydrochloric and phosphoric acids do not affect the action of the indicator, but perchloric acid interferes. The indicator is converted to toluylene red on boiling, so that the titration is best carried out in the cold.

Patent Blue V

This dyestuff, as already mentioned (p. 132), may be used as an acid-base indicator. It is also satisfactory as a reversible redox indicator in the titration of iron-II with ceric sulphate or with potassium permanganate.[33]

The solution should be N with respect to sulphuric acid. If it is necessary to reduce the iron from the iron-III form first, a method must be used which does not involve hydrochloric acid, as this acid produces a dark brownish colour which obscures the indicator colour. Chloride must also be absent, and the indicator cannot be used with dichromate.

The colour change is from yellow in the reduced form to orange-red in the oxidised form. The oxidation potential is 0·78 volt (with reference to the standard hydrogen electrode), which corresponds closely to the equilibrium potential of the ferrous-ferric system (0·77 volt).

Although the colour of the oxidised form fades after a short time, the end-point may be observed by back titration within several minutes of adding excess of oxidising agent.

Indicator solution. 0·1 per cent aqueous.

Use 3–5 drops (sufficient to produce a distinct colour).

Brucine sulphate

Brucine sulphate has been recommended as a redox indicator for the titration of stannous or ferrous ion by dichromate.[34] The colour change is from colourless in the reduced form to red in the oxidised form, and it is claimed that the indicator is preferable to diphenylamine for these titrations.

Indicator solution. Dissolve 1 g. brucine sulphate in 100 ml. concentrated sulphuric acid.

Use 20 drops per titration.

Diphenylcarbazide

This compound has been recommended for the titration of iron with dichromate.[35] The titration procedure is rather lengthy, but the results are claimed to justify this.

To a hydrochloric acid solution of ferrous ion, manganous sulphate solution, ferric sulphate solution and indicator are added, and the solution is titrated with standard dichromate solution under carefully controlled conditions until the end-point is reached. The colour change is from violet through lavender to colourless (for the indicator, the solution after titration being somewhat coloured by the ferric and chromium salts).

Indicator solution. Dissolve 0·1 g. diphenylcarbazide in 30 ml. cold glacial acetic acid. Dilute to 100 ml. with distilled water.

Use 2 ml. indicator solution per titration.

Cacotheline

Cacotheline in the presence of a small amount of ferrous sulphate has been recommended[36] for the direct titration of calcium by sodium oxalate. The solution of the calcium salt should be neutral or acid with acetic acid. On titration with 0·1 N sodium oxalate, calcium oxalate is precipitated. At the end-point the excess sodium oxalate begins to react with ferric ion, favouring reduction of the cacotheline by the ferrous ion, and producing a violet colour which is stable for several minutes.

Magnesium up to at least five times the amount of calcium may be present without interfering.

The same authors have used cacotheline for the titration of iron-III solutions by stannous chloride. The solution of iron-III is adjusted to 2 N towards hydrochloric acid, and is titrated at the boiling point with a 3 per cent solution of stannous chloride in 3 N hydrochloric acid (300 ml. concentrated hydrochloric acid diluted to 1 litre) which has been standardised against potassium iodate. The cacotheline is not reduced until all the iron has been converted to the iron-II form, after which there is a sharp colour change to violet.

Indicator solutions: (a) *Cacotheline.* Saturated aqueous. Use 1 ml.

(b) *Ferrous sulphate.* Saturated aqueous. Use 1 drop in the calcium titration.

[1] T. H. Whitehead and C. C. Wills, *Chem. Reviews*, 1941, **29**, 69.
[2] I. M. Kolthoff, *Analyt. Chem.*, 1949, **21**, 103.
[3] G. H. Walden, L. P. Hammett and R. P. Chapman, *J. Amer. Chem. Soc.*, 1933, **55**, 2649: G. F. Smith and F. P. Richter, *Ind. Eng. Chem. Anal.*, 1944, **16**, 580.
[4] G. F. Smith and F. P. Richter, *loc. cit.*
[5] W. W. Brandt and G. F. Smith, *Analyt. Chem.*, 1949, **21**, 1313.
[6] G. F. Smith, *ibid.*, 1951, **23**, 925.
[7] F. W. Cagle and G. F. Smith, *ibid.*, 1947, **19**, 384.
[8] J. Steigman, N. Birnbaum and S. M. Edmonds, *Ind. Eng. Chem. Anal.*, 1942, **14**, 30.
[9] F. P. Dwyer, J. E. Humpoletz and R. S. Nyholm, *J. Proc. Roy. Soc. N.S. Wales*, 1946, **80**, 212.
[10] D. P. Shcherbov, *Zavod. Lab.*, 1948, **14**, 794.
[11] H. H. Willard and G. D. Manolo, *Analyt. Chem.*, 1947, **19**, 167.
[12] A. V. Kirsanov and V. P. Cherkassov, *Bull. Soc. chim.*, 1936, **3**, 2037.
[13] J. Knop and O. Kubelkova-Knopova, *Z. anal. Chem.*, 1941, **122**, 183.
[14] S. Cohen and R. E. Oesper, *Ind. Eng. Chem. Anal.*, 1936, **8**, 364.
[15] F. Burriel and F. Sierra, *Anal. Fís. Quím.*, 1932, **30**, 441.
[16] L. E. Straka and R. E. Oesper, *Ind. Eng. Chem. Anal.*, 1934, **6**, 405.
[17] R. Belcher and A. J. Nutten, *J.C.S.*, 1951, 547.
[18] R. Belcher, A. J. Nutten and W. I. Stephen, *ibid.*, 1520.
[19] *Idem, ibid.*, 3444.
[20] R. Belcher, A. J. Nutten and W. I. Stephen, *J.C.S.*, 1952, 1269.
[21] I. M. Kolthoff and V. A. Stenger, *Ind. Eng. Chem. Anal.*, 1935, **7**, 79.
[22] R. Belcher, *Anal. Chim. Acta*, 1950, **4**, 468.
[23] *Idem, ibid.*, 1951, **5**, 27.
[24] G. F. Smith and C. S. Wilcox, *Ind. Eng. Chem. Anal.*, 1942, **14**, 49.
[25] R. Belcher and S. J. Clark, *Anal. Chim. Acta*, 1950, **4**, 580.
[26] G. F. Smith and H. H. Bliss, *J. Amer. Chem. Soc.*, 1931, **53**, 2091: G. F. Smith and R. L. May, *Ind. Eng. Chem. Anal.*, 1941, **13**, 460.

[27] F. L. Hahn, *Ind. Eng. Chem. Anal.*, 1942, **14**, 571.
[28] R. Uzel, *Časopis Šeskoslov. Lécarnictva*, 1935, **15**, 143: *C.A.*, 1935, **29**, 6523.
[29] R. Belcher, *Anal. Chim. Acta*, 1949, **3**, 578.
[30] E. Schulek and P. Rósza, *Z. anal. Chem.*, 1939, **116**, 185.
[31] R. Belcher, *Anal. Chim. Acta*, 1951, **5**, 30.
[32] L. S. V. de Bollini, *Anal. farm. bioquim. (Buenos Aires)*, 1947, **18**, 115: *C.A.*, 1949, **43**, 2117.
[33] J. H. Yoe and G. R. Boyd, *Ind. Eng. Chem. Anal.*, 1939, **11**, 492.
[34] S. Miyagi, *J. Soc. Chem. Ind. Japan*, 1933, **36**, Suppl. binding 146: *C.A.*, 1933, **27**, 3418.
[35] H. E. Crossley, *Analyst*, 1936, **61**, 164.
[36] M. L. Kutschment and A. I. Gengrinovitsch, *Zavod. Lab.*, 1945, **11**, 267: *C.A.*, 1946, **40**, 1412.

C. IODOMETRIC INDICATORS

Amylose

The blue colour of "starch-iodide" is caused by the formation of a complex between iodine and amylose. Using amylose itself as indicator, the end-point is more sensitive, in the presence of sufficient potassium iodide, than that obtained with soluble starch.[1] Titrations may be carried out down to 0·001 N iodine.

Indicator solution. 1 per cent aqueous.

Use 2–3 drops of this solution.

Methylene blue

Iodine decolorises methylene blue, forming the hydriodide of tetraiodomethylene blue. Reducing agents such as mercury-II, tin-II, arsenite, hydrosulphide, sulphite and thiosulphate restore the blue colour, so that it is possible[2] to use methylene blue as an iodometric indicator. It is claimed that the indicator is as easy to use as starch, and is more reliable with dilute solutions (0·01 N or less).

Sodium starch glycollate

Sodium starch glycollate is stated to be preferable to starch as an iodometric indicator.[3] The reagent is water soluble and stable, and does not show the drift which occurs with starch in dilute solutions. Since it forms a water-soluble iodine complex it may be added at the beginning of the titration.

Sodium amylose glycollate

This compound has been prepared from amylose[4] in the same way as the starch compound from starch.[3] The preparation is difficult, but it is claimed that the indicator is more sensitive than its starch analogue.

Polyvinyl alcohol

Completely de-acetylated polyvinyl alcohol gives a blue colour with iodine in the same way as starch. A solution of polyvinyl acetate in aqueous ethanol, or alternatively, a solution of polyvinyl alcohol containing at least 10 per cent of residual acetate groups, gives a crimson colour.[5] This is a highly sensitive test for iodine when the alcohol contains about 20 per cent of these residual groups. At very low concentrations of iodine the colour is brownish-yellow and may be observed down to 3×10^{-6} N iodine in 0·0002 per cent potassium iodide solution, or to 10^{-5} N iodine in 0·00004 per cent potassium iodide solution. At these dilutions starch gives no colour.

The indicator is readily soluble in water and the solution is stable. Titrations of iodine with sodium thiosulphate to the discharge of the brownish-yellow colour may readily be carried out in 0·01 N solutions. There is no drift of the end-point, and the indicator may be added at the beginning of the titration.

Indicator solution. 1 per cent aqueous polyvinyl alcohol containing 20 per cent residual acetate groups.

Use 0·5 ml. for the titration of a 5-ml. sample of iodine solution.

[1] L. H. Liggett and H. Diehl, *Anachem. News*, 1946, **6**, 9: *C.A.*, 1946, **40**, 2757.
[2] J. A. Gautier, *Ann. pharm. Franç.*, 1948, **6**, 171.
[3] S. Peat, E. J. Bourne and R. D. Thrower, *Nature*, 1947, **159**, 810.
[4] E. J. Bourne, Private communication.
[5] S. A. Miller and A. Bracken, *J.C.S.*, 1951, 1933.

D. ADSORPTION INDICATORS

Bromothymol blue

Mehrotra[1] has used bromothymol blue in the argentometric titration of thiocyanate ions. As silver nitrate solution is run into the thiocyanate solution, each drop produces a blue colour where it comes into contact with the suspension. Coagulation of the silver thiocyanate begins quite early, but the precipitate and the supernatant suspension remain colourless as long as thiocyanate ions are in excess. At the end-point half a drop of silver solution turns the precipitate blue. The colour change is sharp and reversible for solutions stronger than 0·01 N. The pH must be greater than 7·0.

Indicator solution. 0·1 per cent in ethanol.

Bromocresol purple

Bromocresol purple may be used[1] in the same way as bromothymol blue in the titration of thiocyanate ions with silver solution. The precipitate remains colourless and the supernatant liquid violet while the thiocyanate is present in excess. At the end-point the fluid becomes colourless and simultaneously the coagulated precipitate becomes bluish-green. The end-point is stated to be sharper than that obtained with bromothymol blue. The titration can be carried out in 0·01 N solution, but a sharp end-point is not obtained if the pH is less than 7·0.

Indicator solution. 0·1 per cent in ethanol.

Bromophenol blue

This indicator has been used for the titration of chloride and iodide against silver[2] and for the titration of chloride and bromide against mercurous solution.[3] According to Mehrotra[1] it can be used as an adsorption indicator in the same way as bromothymol blue, though in this case the solution to be titrated may be at a pH as low as 3·0, and a concentration of 0·004 N. In acid solution the supernatant liquid remains colourless if thiocyanate ions are in excess, becoming blue at the equivalence point.

Bromophenol blue may also be used as an adsorption indicator in the titrimetric determination of thallium.[4] On running in thallous nitrate or sulphate to the 0·1 N potassium iodide solution, reddish-yellow thallous iodide begins to coagulate towards the end of the titration, the supernatant suspension remaining violet. A few drops before the end-point the precipitate begins to lighten in colour. The titration should be continued carefully and with shaking after addition of each drop of the thallous solution. At the end-point the precipitate, now greenish-yellow, changes to rich green.

The titration should preferably be carried out in diffuse light, and the indicator added near the end-point. Lead interferes, but may be removed by adding a slight excess of potassium sulphate. No interference occurs, however, in the titration with thallous sulphate, provided that the amount of lead present is less than equivalent to the amount of potassium iodide taken. Chromate and thiocyanate interfere.

The end-point is sharp at pH 4·0–8·0, and when the titrating solutions have a concentration of 0·2–0·04 N.

Indicator solution. 0·1 per cent solution in 80 per cent ethanol. Use 3–4 drops per titration.

Four adsorption indicators for the titration of silver with thiocyanate have been proposed by Uzumasa and Miyake.[5] These are used under slightly alkaline conditions and their characteristics are shown in Table V.

TABLE V

Indicator	Colour change	Minimum effective concentration of medium
Sodium alizarin sulphonate	Yellow to pink	0·033 N
Phenolphthalein	Pink to purple	0·01 N
Cochineal	Purple to green	0·01 N
Phenol red	Red to violet	0·01 N

Congo red

Congo red as an adsorption indicator in halide and thiocyanate determinations may be adsorbed by either positively or negatively charged particles.[6] The colour change is particularly sharp, from blue to red, addition of dextrin rendering it even sharper. It is readily reversible, and titration may be effected from either direction. At the end-point (when adding silver nitrate to chloride) the precipitate changes from blue to pink, and the pink fluid becomes colourless. Chloride may be titrated down to 0·02 N, thiocyanate and bromide to 0·004 N and iodide to 0·001 N. The best pH range for the titration is pH 3·0–5·0.

Indicator solution. 0·2 per cent aqueous recrystallised Congo red. Use 1 drop indicator solution per 10 ml. 0·1 N solution or 20 ml. 0·02 N solution.

Resorcinol-Succinein

Mehrotra and his co-workers used resorcinol-succinein as an adsorption indicator in argentometry.[7] The titration fails in acid solutions, but is possible in neutral or slightly alkaline solutions.

Indicator solution. 0·2 per cent in ethanol.

Use 2 drops per 20 ml. solution.

Mercurochrome

Mercurochrome is elsewhere noted (p. 133) as an acid-base indicator, but Airan and Ghatage[8] have employed it successfully as an indicator in halide-silver titrations. The halide solution should contain 1·5 ml. concentrated sulphuric acid and 1·5 ml. phosphoric acid per 10 ml. Titration is carried out with 0·05 N silver nitrate to a persistent pink colour. It is claimed that the results agree satisfactorily with those obtained using alkali chromate as indicator.

Indicator solution. 0·1 per cent aqueous.

Use 2–3 drops per titration.

Tartrazine

Berry and Durrant[9] used tartrazine as an adsorption indicator in the titration of silver with bromide. The precipitate first adsorbs the dye, and remains yellow as long as silver ions are present in excess. At the end-point the tartrazine is displaced, producing a greenish-yellow colour in the liquid. The first visible green colour appears 0·2–0·25 per cent before the end-point, but an accuracy of 0·1 per cent may be achieved by titrating to a deeper green.

In a later paper Berry[10] describes the use of tartrazine as indicator in the Volhard determination of chloride. The main feature of its use in this connection is that titration can be carried out without either filtration or protection of the precipitate by nitrobenzene. In his experiments, Berry added weighed quantities of silver nitrate to the same volume of approximately 0·1 N hydrochloric acid and completed the titration either with the same stock of hydrochloric acid or with a solution of ammonium thiocyanate.

For the determination of chloride, excess silver nitrate solution is added and the excess silver is determined with either hydrochloric acid or with ammonium thiocyanate. The end-point is indicated when the dye is removed from the precipitate and the supernatant liquid becomes a rich lemon yellow.

Indicator solution. 0·5 per cent aqueous.

Use 3 drops per titration.

Tetraiodophenolsulphonephthalein

After an examination of a range of substances for the argentometric determination of thiocyanate, Mehrotra[11] chose tetraiodophenolsulphonephthalein as the most suitable. If used in neutral solution the precipitate coagulates too early in the titration, and this is accompanied by adsorption of the dye. However, in the presence of nitric acid, a change from yellow to blue is observed, which is very sharp, even at concentrations as low as 0·005 N. The best pH is in the range pH 2·0–4·5 (0·001–0·005 N nitric acid), otherwise a slight excess of silver nitrate must be added to bring about the change.

Indicator solution. 0·2 per cent ethanolic.

Use 2 drops for every 10 ml. thiocyanate solution.

Benzidine-copper complex

It has been found[12] that benzidine, as either the sulphate or the acetate, may be used in the form of a copper complex as an adsorption indicator in silver-bromide or silver-iodide titrations, and that it makes such titrations possible at lower concentrations, and at lower pH than the more usual adsorption indicators, as well as permitting the presence of strongly coloured cations. It is suggested that there is formation of a copper benzidine halide complex which is adsorbed, and that in addition free halogen is formed which oxidises the benzidine to give a deep blue compound, also adsorbed by the silver halide precipitate at the end-point.

In the iodide titration 1 ml. 5 per cent aqueous copper nitrate solution is added to the silver solution, together with the indicator. On titration the solution is kept vigorously agitated, and a very rapid and sharp change from greenish-yellow to greyish-blue is obtained at the end-point.

The presence of the coloured cations of Group III has no effect on the results. Iron modifies the colour change somewhat, but this can be overcome by adding 2 g. solid sodium sulphate, which reduces the dissociation of the benzidine sulphate so that less benzidine is available for oxidation by the ferric ions.

The titration is successful down to 0·001 N solutions, and the pH may be as low as that corresponding to 3 ml. N acetic acid or 2 ml. 0·1 N nitric acid per 20 ml. solution.

Indicator solutions: (a) Benzidine acetate. Warm 0·4 g. benzidine acetate with 100 ml. water, and add glacial acetic acid dropwise (2–3 drops) till solution occurs. Filter the solution.

12

Use 5 ml. of this indicator solution for 0·01 or 0·001 N titrations. of iodide, or for all bromide titrations.

(b) *Benzidine sulphate.* To the acetic acid solution of benzidine acetate obtained as in (a) add an equal volume of a saturated aqueous solution of sodium sulphate.

Use 10 drops of this suspension for N or 0·1 N titrations of iodide.

The titration of bromides is carried out in similar fashion to that of iodides, but the copper nitrate is replaced by copper acetate (1 ml. 5 per cent aqueous solution). At the end-point the colour changes from light yellow to green, this change being followed by coagulation of the silver bromide. Benzidine sulphate was found unsatisfactory for the bromide titration, possibly because of its smaller dissociation. Substituted benzidines have been found to be more sensitive in these titrations than benzidine itself.[12]

[1] R. C. Mehrotra, *Anal. Chim. Acta*, 1949, **3**, 69.
[2] I. M. Kolthoff, *Z. anal. Chem.*, 1927, **71**, 235.
[3] L. von Zombory, *Z. anorg. Chem.*, 1929, **184**, 237: 1933, **215**, 255.
[4] R. C. Mehrotra, *Nature*, 1948, **161**, 242.
[5] Y. Uzumasa and Y. Miyake, *J. Chem. Soc. Japan*, 1933, **54**, 624.
[6] R. C. Mehrotra, *Anal. Chim. Acta*, 1948, **2**, 36.
[7] R. C. Mehrotra, R. D. Tiwari and H. L. Dube, *Curr. Sci.*, 1947, **16**, 119.
[8] J. W. Airan and N. D. Ghatage, *ibid.*, 343.
[9] A. J. Berry and P. J. Durrant, *Analyst*, 1930, **55**, 613.
[10] A. J. Berry, *ibid.*, 1948, **73**, 506.
[11] R. C. Mehrotra, *Anal. Chim. Acta*, 1950, **4**, 38.
[12] F. Sierra and F. Burriel, *Anal. Fís. Quím.*, 1932, **30**, 366: F. Burriel, *ibid.*, 1935, **33**, 692.

E. MISCELLANEOUS INDICATORS

Nickel dimethylglyoxime

In a solution containing nickel ion and cyanide ion the equilibrium

$$\frac{[Ni(CN)_4^=]}{[Ni^{++}]} = K[CN^-]^4,$$

where K is a constant, must exist, and this relation must hold even when excess cyanide ion is present. If a solution of a metallic ion whose cyanide is less dissociated than the nickelocyanide is added, and if the new equilibrium is set up rapidly, nickel ion will be liberated as soon as any excess cyanide which may have been present is used up by the added metal ion. If the solution at the same time contains dimethylglyoxime, the free nickel ion will be complexed immediately to the red organometallic complex.

The correct conditions are produced when an ethanolic solution of potassium cyanide is boiled with an excess of solid nickel dimethylglyoxime and filtered, the resulting solution being in effect a solution of potassium nickelocyanide together with free dimethylglyoxime. This solution may be used as indicator in the titration of cyanide with certain heavy metal solutions.[1]

With silver and copper, particularly the former, the equilibrium is set up rapidly enough to permit titration in the normal fashion. Indicator solution is added to standard potassium cyanide in a conical flask. The silver or copper solution is run in until a pink colour appears at the end-point, the last few drops being added at intervals of not less than 30 seconds, with thorough shaking.

If mercury salts are to be determined the slowness of the reaction makes it preferable to add the solution to a measured excess of cyanide, and to allow this to stand for some time before completing the titration with standard silver nitrate solution.

The potassium cyanide solution should be standardised frequently, since 0·1 N solution loses in strength at the rate of about 0·3 per cent per day.

Indicator solution. Dilute a 5 per cent solution of potassium cyanide with an equal volume of ethanol. Add excess solid nickel dimethylglyoxime, and boil. Allow to stand for some time, and filter. Dilute to one-tenth strength with water. Filter the indicator solution at intervals if a deposit appears.

Use a few drops of the diluted solution per titration.

The same reaction has been employed by Burriel and Pino[2] for the titration of cyanide by silver, the principal difference being that the indicator is prepared *in situ*. To the cyanide solution (10 ml., diluted with water to about 90 ml.) is added 0·5 to 1 ml. of a 1 per cent ethanolic solution of dimethylglyoxime and 3 drops of 1 per cent nickel sulphate heptahydrate solution. The cyanide prevents formation of the red complex. Silver nitrate solution is then added until a white cloudiness appears. The solution is shaken thoroughly, and addition of the silver nitrate is continued, drop by drop, until the suspension turns red throughout. The titration is not reversible.

The same authors have applied the principle to the titration of an acid with a base, utilising the solubility of the complex in acid solution. To the acid solution, so adjusted that the final volume will be about 100 ml., are added 4–5 drops nickel sulphate solution and 0·5–1·0 ml. dimethylglyoxime solution. Titration is carried out with the alkali. A few drops before the neutral point the solution becomes markedly yellow. The final end-point is marked by a pink colour which does not disappear on standing. Although the colour change is quite similar to that with phenolphthalein, the dimethyl-glyoxime indicator has the advantage that it is not CO_2-sensitive. It is, however, not reversible, because of the slowness with which the precipitate, once formed, is taken up into solution again by acid.

p-Dimethylaminobenzylidenerhodanine

This substance has been proposed as a suitable indicator for the titration of cyanide with silver solution.[3] An acetone solution of the compound is yellow, while the silver complex is red violet. The titration is carried out in alkaline solution, and since coloured precipitates are formed by a number of metals in alkaline solution, no metal other than the alkali metals should be present.

To the cyanide solution in a flask is added 10 ml. 10 per cent sodium hydroxide solution followed by the indicator. Silver nitrate solution is added until the solution shows a sharp colour change from pale yellow to red violet. The indicator is stated to be preferable to potassium iodide, particularly for the titration of dilute (0·02–0·04 N) solutions.

Indicator solution. 0·02 per cent in acetone.

Use 3 drops per titration.

o-Dianisidine

This substance has been recommended[4] as an indicator for the titration of zinc with potassium ferrocyanide. It actually functions in a redox system, since potassium ferrocyanide normally contains a trace of ferricyanide sufficient to give a red colour with the reagent in acid solution. At the end-point the excess of ferrocyanide lowers the oxidation potential sufficiently to discharge the red colour.

To the zinc solution 10 g. ammonium chloride and 5 ml. concentrated sulphuric acid are added, and titration with 0·05 M potassium ferrocyanide is begun. Towards the end of the titration sufficient indicator is added to colour the solution red-brown. The titration is continued dropwise to the end-point, which is marked by a change to colourless or pale green. If the potassium ferrocyanide contains insufficient ferricyanide as impurity, 0·1 per cent potassium ferri-cyanide solution may be added dropwise to obtain the required red-brown colour.

Metals forming insoluble ferrocyanides interfere.

Indicator solution. Mix 0·1 g. *o*-dianisidine with 0·5 ml. concentrated sulphuric acid, and dilute to 100 ml. with distilled water.

Diphenylcarbazide

This compound has been recommended[5] for the titration of chloride with mercuric nitrate, and the method has been further improved.[6]

The volume of chloride solution is adjusted so that the final volume of solution will be about 80–100 ml. The indicator is added, and if necessary the solution is neutralised with sodium hydroxide solution to an orange colour. The solution is then acidified with 4 ml. 0·2 N nitric acid, and titration is carried out with 0·1 N mercuric nitrate solution.

About 5 drops before the end-point a pink-violet colour begins to develop. At the end-point this changes sharply to deep blue-violet.

Indicator solution. Saturated diphenylcarbazide in 95 per cent ethanol. This solution gradually turns red on standing, but its action does not appear to be affected.

Use 5 drops per titration.

Diphenylcarbazone

Roberts[6] has also used diphenylcarbazone in the mercury-chloride titration, down to 0·025 N. The chloride solution is adjusted to give a final volume of about 65 ml. It is titrated with alkali to the full blue colour of bromophenol blue, and is then acidified with 4 ml. 0·2 N nitric acid. Titration is carried out with 0·025 N mercuric nitrate solution to a definite pink colour of the diphenylcarbazone indicator, preferably using a daylight lamp, although the yellow of the bromophenol blue does not interfere materially with the end-point. At high dilution a blank determination should be carried out, since the blank may be as much as 0·07 ml. in 0·025 N solution.

Indicator solutions: (a) Bromophenol blue. See p. 158.

Use 2 drops.

(b) Diphenylcarbazone. Saturated ethanolic.

Use 5 drops.

Diphenylcarbazone in conjunction with a trace of a mercury salt has also been proposed[7] as indicator for silver-halide or silver-thiocyanate titrations in coloured solutions. In a slightly acid medium insufficient mercuric ions are produced to react with the diphenyl-carbazone, but on addition of excess silver sufficient mercuric ions are liberated to produce a colour.

For the titration of chloride a mercuric chloride-diphenylcarbazone indicator is used. Sufficient acid should be present to correspond to not more than 1 ml. concentrated nitric acid in a final volume of 100 ml. Ether is added, and the titration is carried out until the pale brown ether layer changes sharply through purple to reddish-brown. The same procedure is used for bromide titrations, but it is not necessary to extract the complex since adsorption does not interfere, and consequently ether may or may not be used as preferred.

For iodide titrations mercuric iodide-diphenylcarbazone is used, and ether is not necessary. The colour change is observed on the surface of the precipitate. Mercury thiocyanate-diphenylcarbazone is used for thiocyanate titrations, with or without ether.

Indicator solutions: (*a*) Dissolve 1 g. mercuric chloride and 0·5 g. diphenylcarbazone in 100 ml. acetone.

(*b*) Dissolve 0·5 g. mercuric iodide and 0·5 g. diphenylcarbazone in 100 ml. acetone.

(*c*) Dissolve 1 g. mercuric thiocyanate and 0·5 g. diphenylcarbazone in 100 ml. acetone.

Use 1 ml. of the appropriate indicator solution per titration.

Diphenylcarbazide disodium disulphonate, when oxidised immediately before titration to the diphenylcarbazone compound, has proved successful as an indicator in the mercurimetric titration of chloride (p. 220).

Ammonium sulphatocerate-starch

Silver solutions may be titrated with potassium iodide using cerate in the presence of starch as a means of indicating the end-point.[8]

The silver solution, approximately 0·1 N, is placed in a 200-ml. beaker and diluted to approximately 110 ml., adjusting the acidity towards sulphuric acid to about 0·2–0·3 N. Starch solution and ammonium sulphatocerate solution are added. On addition of 0·1 N potassium iodide oxidation to iodine is not permanent until the end-point is reached, so that any blue colour which appears is transient and disappears on stirring. At the end-point the solution turns a permanent blue-green. A blank correction should be applied.

The titration gives good results in the presence of ferric or cupric ions.
Indicator solutions: (*a*) *Starch solution.* 0·5 per cent aqueous.
Use 3 ml. per titration.
(*b*) *Ammonium sulphatocerate.* Approximately 0·1 N.
Use 0·1 ml. per titration.

Sodium-2 : 5-cresotate

This compound reacts with uranyl acetate to form a red complex.
The reaction may be utilised in the titration of phosphate by uranyl
acetate solution.[9]
The phosphate solution (2 ml.) is placed in a beaker, and 10 ml.
distilled water is added, followed by 2 ml. of a solution of the indicator, 1 ml. hexamethylenetetramine solution, and 0·5 g. potassium
nitrate. Uranyl acetate solution is added from a microburette, drop
by drop, until a turbidity forms. This is coagulated by boiling, and
uranyl acetate addition is continued until a red-brown colour appears.
A parallel procedure is applied to a standard phosphate solution containing 2·5 mg. P_2O_5 per ml. The volumes are equalised if necessary
by adding water, and the titration is completed to a colour match.

Siloxene

Siloxene is the name applied to a derivative of calcium silicide,
first prepared by Wöhler,[10] and probably of variable composition,
but assigned the formula $(Si_6H_6O_3)_n$. In acid solution it reacts with
strong oxidising agents, with the immediate production of intense
chemiluminescence. This reaction has long been used as a qualitative
test for calcium silicide.
The substance has been recommended[11] as an indicator for use
with strong oxidising agents, particularly in coloured solutions where
a more conventional indicator would be masked. Potential measure-
ments at the stoicheiometric point show an oxidation potential of
1·17 $\pm$ 0·03 volt (referred to the normal hydrogen electrode), while
sufficient light to indicate the end-point is emitted about 0·09 volt
above this value.
In particular, the indicator has been found satisfactory in titrations
of ferrous solutions by ceric sulphate solution, both in the presence
and absence of cobalt.
A suitable amount of indicator is 100 mg. of the solid, added to
25 ml. reducing (ferrous) solution, which has been adjusted to a pH
of 2·0 or less.

The titration is carried out in a dark room with standard 0·1 M ceric sulphate solution. On each addition a local spot of bright light is produced, which disappears on shaking until the end-point is reached, when the bright outline of the whole liquid is visible.

The indicator blank is of the order of 0·04 ml., and for many purposes may be disregarded.

Synthesis of siloxene. Place 5 g. finely powdered technical calcium silicide in an 800-ml. beaker in a fume cupboard. Add 50 ml. concentrated hydrochloric acid, and stir for several minutes with a long stirring rod. A vigorous reaction takes place, and much hydrogen chloride is evolved. When the reaction has subsided, add a further 25 ml. concentrated hydrochloric acid and boil gently with stirring for 5 minutes. Add 150 ml. water and boil for a further 5 minutes. Decant the yellow suspension of siloxene from unchanged calcium silicide through a filter paper on a Büchner funnel. Wash the siloxene with water, with 95 per cent ethanol and with ether. Spread the yellow product on a porous plate to dry. Store in an amber glass bottle, away from light.

Yield about 5 g. Time of preparation about 20 minutes.

The material prepared in this way is stable for only a few days when dry, but for longer if stored under dilute hydrochloric acid.

It is not spontaneously inflammable or explosive if this method of preparation is followed. However, samples of technical calcium silicide vary very greatly, and it is advisable to carry out tests with small amounts of material from a fresh batch before proceeding to a full-scale preparation, since, in the present authors' experience, the reaction of some specimens may be extremely vigorous, presumably owing to simultaneous evolution of hydrogen and spontaneously inflammable silicon hydrides.

Luminol

Luminol (5-amino-2 : 3-dihydro-1 : 4-phthalazinedione)

is oxidised in alkaline solution by hydrogen peroxide, in the presence of a suitable catalyst, with the emission of strong chemiluminescence. The light emission is maximum at pH 12·0–12·7, but a perceptible light is visible when sufficient sodium hydroxide is added to an acid solution to produce a pH of 7·0. Kenny and Kurtz[12] have recommended luminol for use as an indicator in titrating an acid with a base in highly coloured solutions. Sufficient gentian violet to obscure completely the colour change of ordinary acid-base indicators has no effect.

The oxidation is irreversible, so that titration must be from the acid to the alkaline side. If it is desired to titrate a base, a known excess of acid must be added, followed by back titration.

The catalyst recommended, which must be stable in acid and alkaline solution, is haemoglobin. The amounts of luminol, hydrogen peroxide and haemoglobin present may vary over wide ranges, but those recommended are about 0·1 per cent of luminol, 0·3 per cent of hydrogen peroxide and 0·1 per cent of haemoglobin (at the beginning of the titration).

Indicator. Add 40 mg. luminol, 6 ml. 3 per cent hydrogen peroxide and 30 mg. haemoglobin.

[1] A. R. Ubbelohde, *Analyst*, 1934, **59**, 339.
[2] F. Burriel and F. Pino, *Anal. Fís. Quím.*, 1949, B, **45**, 43.
[3] J. A. Ryan and G. W. Culshaw, *Analyst*, 1944, **69**, 370.
[4] H. F. Frost, *ibid.*, 1943, **68**, 51.
[5] J. V. Dubsky and J. Trtilek, *Mikrochem. Mikrochim. Acta*, 1933, **12**, 215; 1934, **5**, 302; *Z. anal. Chem.*, 1933, **93**, 345.
[6] I. Roberts, *Ind. Eng. Chem. Anal.*, 1936, **8**, 365.
[7] C. H. R. Gentry and L. G. Sherrington, *Analyst*, 1945, **70**, 460.
[8] A. Bloom and W. M. McNabb, *Ind. Eng. Chem. Anal.*, 1936, **8**, 167.
[9] B. R. Repman, *Lab. Prakt. (U.S.S.R.)*, 1941, **16**, No. 1, 27: *C.A.*, 1941, **35**, 4305.
[10] F. Wöhler, *Annalen*, 1863, **127**, 263.
[11] F. Kenny and R. B. Kurtz, *Analyt. Chem.*, 1950, **22**, 693.
[12] *Idem, ibid.*, 1951, **23**, 339.

CHAPTER VI

TITRANTS

1. SODIUM CHLORITE

SODIUM CHLORITE liberates iodine from potassium iodide solutions in the presence of acid, and this reaction may be used for the determination of the strength of chlorite solutions. Bray,[1] who first described the determination of chlorite by thiosulphate, recommended a wait of 5 minutes before titrating in the presence of acetic acid. Schulek and Endroï[2] preferred phosphoric acid, allowing 5 minutes between mixing the solutions and titrating. The use of acetic acid was recommended by Jackson and Parsons.[3] Kolthoff and Sandell[4] claimed that the reaction in the presence of acetic acid is too slow, and recommended the use of sulphuric acid instead, when the titration can be carried out immediately after addition of the acid.

Brown[5] has investigated all the procedures, and finds that in the presence of sulphuric acid or phosphoric acid immediate titration is possible, but with acetic acid a 5-minute interval is necessary before titrating. Brown also stresses the necessity for adding the potassium iodide solution to the chlorite first, as addition of the acid first will produce chlorine dioxide and give rise to low results.

$$ClO_2^- + 4I^- + 4H^+ \longrightarrow Cl^- + 2I_2 + 2H_2O.$$

Sodium chlorite is available commercially as a white, finely divided crystalline material. The salt is slightly hygroscopic. Freshly prepared solutions are clear and colourless, but more concentrated solutions gradually turn yellow and then yellow-green if allowed to stand exposed to light.

The reagent may be used as an oxidising titrant for the determination of sulphurous acid, bisulphites and sulphites.[6] The oxidation is smooth and easily controlled. It is believed to proceed as follows:

$$ClO_2^- + 4H^+ + 4e^- \longrightarrow Cl^- + 2H_2O.$$

Parsons[7] considers this reaction too slow for practical purposes. It is therefore accelerated by the addition of small amounts of potassium iodide, and the reactions then appear to be liberation of iodine according to the reaction already quoted, followed by reduction of the iodine to iodide by the sulphite:

$$2H_2O + 2I_2 + 2SO_3^{=} \longrightarrow 2SO_4^{=} + 4I^- + 4H^+.$$

In both cases the equivalent weight is one-fourth the molecular weight.

Solutions containing lignin derivatives, sugars and other organic matter may be titrated without interference.[8]

Sodium chlorite reacts quantitatively with arsenite in sodium bicarbonate solution, in the presence of osmic acid as catalyst:

$$2AsO_3^{=} + ClO_2^{-} \longrightarrow 2AsO_4^{=} + Cl^-.$$

This reaction has also been investigated by Brown.[5] No satisfactory indicator was found for the direct titration, although the osmic acid itself may act as indicator if the titration is carried out very slowly and with great care. However, excess standard sodium arsenite may be added to chlorite, and the excess determined by back titration with standard iodine, using starch as indicator. Titration in the reverse direction, by addition of excess chlorite solution, followed by addition of potassium iodide and titration of the iodine with sodium thiosulphate, proved impossible, since iodine is liberated by arsenate in acid solution.

Procedures

Standardisation of sodium chlorite. Prepare a solution of 0·1 N sodium chlorite by weighing out the appropriate amount and dissolving in water (2·261 g. per l.). Add an accurately measured amount of this solution to a stoppered conical flask containing 75 ml. distilled water, 15 ml. 10 per cent potassium iodide solution, and 15 ml. 30 per cent acetic acid. Stopper the flask, and leave in the dark for 5 minutes. Titrate the liberated iodine with standard sodium thiosulphate solution, using 5 ml. 1 per cent starch indicator near the end-point.

Store the chlorite solution in a black-painted bottle to exclude light. In these conditions the solution is stable for several months.

Determination of sulphite. Transfer the sulphite solution to a conical flask containing 100 ml. distilled water, 15 ml. 10 per cent potassium iodide solution, 15 ml. 30 per cent acetic acid and 5 ml. 1 per cent starch solution. Titrate rapidly with the sodium chlorite until near the end-point, and then dropwise until the blue colour persists. It is preferable to avoid shaking the solution until most of the sodium chlorite has been added.

$$1 \text{ ml. } 0 \cdot 1 \text{ N chlorite} \equiv 4 \cdot 003 \text{ mg. } SO_3^{=}.$$

[1] W. Bray, *Z. Phys. Chem.*, 1905, **54**, 576.
[2] E. Schulek and P. Endroï, *Anal. Chim. Acta*, 1951, **5**, 368.
[3] D. T. Jackson and J. L. Parsons, *Ind. Eng. Chem. Anal.*, 1937, **9**, 14.
[4] I. M. Kolthoff and E. B. Sandell, *Textbook of Quantitative Inorganic Analysis*, 3rd Edn., New York and London, 1952, p. 597.
[5] E. G. Brown, *Anal. Chim. Acta*, 1952, **7**, 494.
[6] G. R. Levi, *Gazzetta*, 1933, **52**, 58; D. T. Jackson and J. L. Parsons, *loc. cit.*
[7] J. L. Parsons, *Ind. Eng. Chem. Anal.*, 1937, **9**, 250.
[8] G. R. Levi, *ibid.*, 250.

2. POTASSIUM IODATE

In the presence of a large excess of hydrochloric acid potassium iodate reacts with various oxidisable substances by reactions which are, in effect, the following:

$$IO_3^- \longrightarrow ICl$$
$$\text{or}$$
$$I^{5+} + 4e^- \longrightarrow I^+,$$

so that the equivalent of potassium iodate in these reactions is one-quarter of the molecular weight.

In practice the reaction takes place in at least two stages,

$$I^{5+} + 5e^- \longrightarrow I$$
$$\text{and}$$
$$I \longrightarrow I^+ + e^-$$

which may be represented:

$$4IO_3^- + 24H^+ + 20e^- \longrightarrow 2I_2 + 12H_2O \tag{1}$$

$$IO_3^- + 2I_2 + 6H^+ \longrightarrow 5I^+ + 3H_2O \tag{2}$$

the overall stoicheiometric equation being

$$IO_3^- + 6H^+ + 4e^- \longrightarrow I^+ + 3H_2O.$$

This use of potassium iodate as an oxidimetric titrant in high concentrations of hydrochloric acid has been practised for some considerable time. The method was originally developed by Andrews[1] and extended by Jamieson.[2] Carbon tetrachloride was used as an immiscible solvent to indicate the end-point through the complete disappearance of the violet iodine colour. This occurs on final oxidation of the iodine liberated in reaction (1) by the further addition of iodate, as indicated in reaction (2).

The distillation of arsenic as the trichloride and titration with potassium iodate has long been a recognised method for the determination of arsenic, although it has not been widely applied. The reaction

$$2AsCl_3 + KIO_3 + 6HCl \longrightarrow 2AsCl_5 + ICl + KCl + 3H_2O$$

is claimed to be stoicheiometric for concentrations of hydrochloric acid corresponding to the range 3–6 N, the reaction being catalysed by ICl. The end-point is sharp, taking place within one drop of 0·005 M potassium iodate.

Using this reaction as a finish, a rapid method for the separation of arsenic from glass and similar substances and its subsequent determination has been worked out.[3] The sample of glass is fused with solid caustic alkali, the low temperature used avoiding the necessity for adding sodium nitrate to prevent loss of arsenic. The melt is then dissolved in hydrochloric acid. The conditions are so arranged that silica is not precipitated at first, and the gelatinous precipitate which subsequently forms gives no trouble in the distillation process. The arsenic is reduced by hydrazine, potassium bromide acting as catalyst. Arsenic trichloride is distilled off in two stages with concentrated hydrochloric acid. The distillate is titrated directly with standard potassium iodate solution. The 0·005 N solution used is conveniently prepared by dilution from a stronger stock solution, say 0·1–0·02 N. It is stable, and does not require standardisation.

The iodine monochloride necessary to catalyse the reaction is most conveniently obtained by adding 0·02 N arsenic trioxide solution at the beginning of the titration, subsequently making the necessary correction for the amount of added arsenic. In this way the necessary catalyst is prepared *in situ*. However, if more than 2 mg. arsenic trioxide is already present in the solution being titrated, this prior addition is unnecessary. The end-point is observed by concentration of the liberated iodine in a carbon tetrachloride layer.

The distillation method employed separates arsenic from all other common elements. In particular, amounts of antimony up to double the amount of the arsenic, and fluorine do not interfere. The method has been applied successfully to lead glasses and borosilicate glasses with a high arsenic content.

The only substances likely to cause interference would be those likely to prevent evolution of arsenic trichloride, such as sulphur, sulphides or thiosulphates. If all reagent and standard solutions are already prepared, the time required for a complete determination is 2 hours.

The usefulness of potassium iodate has been much extended by the discovery[4] that certain dyestuffs may be used as internal indicators which are destroyed at the end-point (p. 150). This offers the obvious advantage of a homogeneous medium. In addition, these indicators allow antimony to be titrated in the presence of tartrate, a titration in which the Andrews-Jamieson method fails.

Three indicators are recommended by Smith and Wilcox. Their properties are given in Table VI.

TABLE VI

Indicator	British Colour Index Number	Colour change
Naphthol Blue Black	246	Green to faint pink
Brilliant Ponceaux 5 R	185	Orange to colourless
Amaranth	184	Red to colourless

It was found that 1 ml. of an aqueous 0·2 per cent solution of the indicator is destroyed by 0·05 ml. of 0·1 N potassium iodate. The indicator blank is therefore negligible, since less than 0·5 ml. indicator solution is used in each titration. To counteract the tendency of the indicator to fade, the addition is delayed until the equivalence point is nearly reached.

The Andrews method has been applied by Heisig[5] to the determination of iron, and is particularly suited to this determination when organic matter is present. The method works equally well when the Smith-Wilcox indicators are substituted for the organic solvent.

In the determination it is advisable to standardise the potassium iodate against iron as a reference. Smith and Wilcox found a value of 0·1009 N for a solution standardised against ferrous sulphate, and this same solution, when evaluated against potassium permanganate and ceric sulphate, showed satisfactory agreement. The same solution of potassium iodate, however, gave a value of 0·1000 N when standardised against pure arsenious oxide, hence low results may be obtained if the latter is used as a standard of reference. The same authors titrated samples containing about 0·15 g. iron in the presence of 2 g. tartaric acid, 2 g. citric acid, 1 g. sodium oxalate, 10 ml. 95 per cent ethanol or 10 ml. glycerol. Excellent results were obtained in every case, indicating the usefulness of the method for the titration of iron in the presence of organic matter.

Thallium, thiosulphate, thiocyanate, sulphurous acid, hydrogen peroxide, hydrazine and phenylhydrazine may also be titrated by iodate. Phenylhydrazine cannot normally be determined by the Andrews-Jamieson procedure, since it appears to contain some impurity which is extracted by the immiscible solvent, obscuring the colour change. The internal indicators, however, may be used satisfactorily.

Procedures

Determination of arsenic in glass. Fuse 5 g. pure sodium hydroxide in a silver or nickel crucible, and when it is fairly cool add a 1-g. sample of the finely ground glass. If the arsenic trioxide content is above 2 per cent use 3 g. sodium hydroxide and 0·5 g. sample. Fuse cautiously, and after the initial violent reaction heat either until the melt is completely liquid or until it is dull red, whichever happens first. Cool, and extract with 30–40 ml. water, thus breaking up the melt. Wash the extract into a beaker, and cool the extract and washings. Prepare a distilling apparatus as follows: Fit the cork of a 500-ml. distilling flask with a thermometer whose bulb is just inside the body of the flask when the apparatus is assembled, and with an 80-ml. dropping funnel whose stem reaches almost to the bottom of the flask. Place the flask on a hole cut in a piece of asbestos board so that it fits the flask at the 50-ml. level. In this way, while the flask is heated with a naked Bunsen flame, superheating of the walls of the distilling flask, and consequent adverse effect on the reducing action of the hydrazine, are avoided. Alter the side-arm of the distilling flask so that it is inclined upwards instead of downwards, and bend it at the end so as to fit, preferably through a ground-glass joint, into a vertical condenser with a jacket 12 inches long and a delivery tube, whose end is cut obliquely, which projects a further 6 inches. This delivery tube should be adjusted so that the end dips below liquid level in the receiver.

In the distilling flask place 20 ml. concentrated hydrochloric acid (36 per cent, or preferably 39 per cent, and arsenic-free), 30 ml. water and a boiling-stone. Pour in the extract from the beaker, keeping the contents of the flask constantly agitated. Give the crucible a final rinse with 6 N hydrochloric acid, adding this rinsing to the distilling flask. Add 10 ml. 5 per cent hydrazine hydrate solution acidified with hydrochloric acid, or a corresponding amount of hydrazine hydrochloride, and 0·25 g. potassium bromide. The total amount of liquid in the distilling flask should now be about 120 ml. Assemble the distilling apparatus, and distil as rapidly as possible (while ensuring a cool distillate) until the bulk is reduced to one-half. Keep the end of the condenser dipping below the liquid in the receiver throughout the distillation.

Add concentrated hydrochloric acid from the dropping funnel until the temperature shown by the thermometer rises to about 110° C., and then add a further 50 ml. as rapidly as possible. During this process allow a little air to be drawn in from time to time through the dropping funnel to avoid sucking back of the distillate. The temperature should now fall to 105° C. or lower. Distil until the bulk of liquid in the flask is about 50 ml.

Change the receiver for one containing 50 ml. water, add a further 50 ml. concentrated hydrochloric acid to the distilling flask, and distil once more to a bulk of 50 ml.

Cool the first distillate to 20°–25° C. and transfer it to a 500-ml. graduated flask. Add 5–10 ml. carbon tetrachloride (free from oxidising and reducing substances) and follow this by 2·00 ml. 0·02 N arsenic trioxide solution added from a 10-ml. burette graduated in 0·02 ml. Titrate with 0·005 N potassium iodate from a similar burette, adding the solution slowly at first, and shaking constantly and vigorously. The carbon tetrachloride layer, best observed by inverting the stoppered flask so that it is contained in the neck, becomes violet, and the aqueous layer is brown or yellow depending on the arsenic content of the sample. The end-point is taken when the colour just disappears from the carbon tetrachloride layer.

Add the second distillate to the flask, and continue titrating to the new end-point. If there is any considerable difference between the first and second end-points, carry out a third distillation in the same manner as the second, and add and titrate this distillate also. The arsenic content is obtained from the total titration figure, corrected for the standard arsenic solution added at the beginning of the titration.

1 ml. 0·005 N potassium iodate$\equiv$0·247 mg. As_2O_3.
$\equiv$0·187 mg. As.

General determination of arsenic. Transfer 25 ml. trivalent arsenic solution to a 400-ml. beaker and add sufficient hydrochloric acid and water to make 100 ml. 4·4 M hydrochloric acid. Titrate with 0·1 N potassium iodate solution until almost all of the iodine colour has disappeared, and then add a few drops of one of the Smith and Wilcox indicators (pp. 174, 179) and continue the titration, stirring well between each addition, until the appropriate colour change takes place. The change with Naphthol Blue Black is slower than that with the other two indicators.

1 ml. 0·1 N potassium iodate$\equiv$3·745 mg. As.

Determination of antimony in potassium antimonyl tartrate. Transfer an aliquot of the solution to a 400-ml. beaker, add 50 ml. concentrated hydrochloric acid (s.g. 1·19) and dilute to 100 ml. (6·2 N in hydrochloric acid). Titrate the solution with 0·1 N potassium iodate using one of the three indicators, as in the method for arsenic.

Amaranth is the most satisfactory indicator in this titration. The colour change with Brilliant Ponceaux 5 B is not satisfactory, and Naphthol Blue Black is slower in reaction. In the original Andrews-Jamieson method the violet colour fails to develop in the organic solvent when tartrate is present, hence the new procedure is more suited for the determination of antimony since it is usually desirable to add tartrate to prevent hydrolysis.

1 ml. 0·1 N potassium iodate$\equiv$6·088 mg. Sb.

Determination of ferrous iron. Transfer an aliquot of the solution to be analysed to a 350-ml. conical flask, add 50 ml. concentrated hydrochloric acid, and dilute to 100 ml. Add 10 ml. iodine monochloride solution (prepared by dissolving 0·279 g. potassium iodide and 0·172 g. potassium iodate in 250 ml. water and adding 250 ml. hydrochloric acid s.g. 1·19). Titrate with 0·1 N potassium iodate solution, and add the indicator when the end-point is being approached.

1 ml. 0·1 N potassium iodate≡5·584 mg. Fe.

Determination of purity of iron. Dissolve the iron in 10 ml. concentrated hydrochloric acid, cool, and pass through a Walden silver reductor. Add 40 ml. concentrated hydrochloric acid, dilute to 100 ml. with water, and titrate as before.

Determination of thallium. Transfer an aliquot of the thallium solution to a 250-ml. beaker, add 40 ml. concentrated hydrochloric acid, and dilute to 100 ml. with water. Titrate with 0·1 N potassium iodate until the iodine has almost disappeared, and then add 0·5 ml. amaranth and continue the titration to the destruction of the indicator colour.

1 ml. 0·1 N potassium iodate≡1·022 mg. Tl.

Determination of thiosulphate. Transfer a suitable aliquot of the solution to a conical flask, add 50 ml. concentrated hydrochloric acid, and dilute to 100 ml. Titrate with 0·1 N potassium iodate solution until practically all the liberated iodine has been oxidised to iodine monochloride. Add 0·5 ml. indicator solution and continue the titration to the destruction of the colour. Any one of the three indicators may be used, but amaranth is best.

1 ml. 0·1 N potassium iodate≡1·402 mg. $S_2O_3^=$.

Determination of thiocyanate. Transfer a suitable aliquot to a conical flask, add 50 ml. concentrated hydrochloric acid, and dilute to 100 ml. with water. Begin the titration with 0·1 N potassium iodate immediately to prevent air oxidation at this high acidity, and when all but the last trace of iodine has been oxidised add 0·5 ml. amaranth solution and complete the titration.

1 ml. 0·1 N potassium iodate≡2·901 mg. CNS^-.

Determination of sulphurous acid. Transfer a suitable aliquot of the solution of sulphurous acid to a conical flask, add 40 ml. concentrated hydrochloric acid, and dilute to 100 ml. Titrate with 0·1 N potassium iodate solution as in previous determinations, using any one of the three indicators. Amaranth gives the sharpest end-point in this titration.

$$1 \text{ ml. } 0.1 \text{ N potassium iodate} \equiv 4.003 \text{ mg. } SO_3^=.$$

Determination of hydrogen peroxide. Transfer 30 ml. 0·1 N sodium arsenite to a glass-stoppered conical flask, and add 10 ml. 10 per cent sodium hydroxide solution. Add slowly from a burette with constant agitation 10·0 ml. hydrogen peroxide solution (approximately 0·18 N) and allow to stand for 2 minutes to permit the reaction to go to completion. Add 40 ml. concentrated hydrochloric acid, and shake the flask and contents vigorously. Titrate the excess arsenite with 0·1 N potassium iodate as in previous determinations, using any one of the three indicators. Although all three are satisfactory, amaranth gives the sharpest end-point.

$$1 \text{ ml. } 0.1 \text{ N sodium arsenite} \equiv 1.701 \text{ mg. } H_2O_2.$$

Determination of hydrazine. Transfer a suitable aliquot of hydrazine sulphate solution to a 500-ml. glass stoppered conical flask, add 40 ml. concentrated hydrochloric acid, and dilute to 100 ml. with water. Titrate as in the previous methods with 0·1 N potassium iodate solution, using either amaranth or Brilliant Ponceaux 5 R.

$$1 \text{ ml. } 0.1 \text{ N potassium iodate} \equiv 1.501 \text{ mg. } N_2H_4.$$

Determination of phenylhydrazine. Proceed in the same way as in the determination of hydrazine, but a beaker may be employed for the titration. Brilliant Ponceaux 5 R is the best indicator to use in the determination.

$$1 \text{ ml. } 0.1 \text{ N potassium iodate} \equiv 5.407 \text{ mg. } C_6H_5.N_2H_3.$$

TABLE VII

Determination	B.C.I. No. of recommended indicators for iodate titrations
Arsenic	184, 185
Antimony	184
Iron	184, 185, 246
Thallium	184
Thiosulphate	184
Thiocyanate	184
Sulphurous acid	184

TABLE VII—*continued*

Determination	B.C.I. No. of recommended indicators for iodate titrations
Hydrogen peroxide	184
Hydrazine	184, 185
Phenylhydrazine	185

[1] L. W. Andrews, *J. Amer. Chem. Soc.*, 1903, **25**, 756.
[2] G. S. Jamieson, *Ind. Eng. Chem.*, 1911, **3**, 250: 1916, **8**, 500: 1918, **10**, 290: 1919, **11**, 296: *Amer. J. Sci.*, 1912, **33**, 349, 352: 1914, **38**, 166: *J. Amer. Chem. Soc.*, 1917, **39**, 246: 1918, **40**, 1036.
[3] H. N. Wilson, *Analyst*, 1943, **68**, 361.
[4] G. F. Smith and C. S. Wilcox, *Ind. Eng. Chem. Anal.*, 1942, **14**, 49.
[5] G. B. Heisig, *J. Amer. Chem. Soc.*, 1928, **50**, 1687.

3. CALCIUM HYPOCHLORITE

Kolthoff and Stenger[1] have recommended calcium hypochlorite as a standard oxidising agent. The solution is prepared from the commercial product, and can be kept in dark glass bottles for a long time without appreciable decomposition. Sodium hypochlorite will serve equally well as an oxidising agent.

The reagent is of special advantage for oxidation in a neutral or alkaline medium. It reacts fairly rapidly with various reducing substances which are not oxidised, or are oxidised only slowly, by permanganate or ceric sulphate. Since hypobromite reacts more rapidly than hypochlorite in many reactions, it may be used with advantage in place of the latter; this is best done by adding bromide to the solution and titrating with hypochlorite, when the following reaction occurs:

$$OCl^- + Br^- \longrightarrow OBr^- + Cl^-.$$

The hypochlorite is usually added in excess, and the amount unconsumed in the reaction is then determined either by adding potassium iodide and acid and titrating the liberated iodine with thiosulphate solution, or by adding an excess of standard arsenious oxide solution and titrating the excess with hypochlorite using Bordeaux B as an irreversible indicator.

In the latter case the reaction which occurs is

$$2OCl^- + As_2O_3 \longrightarrow 2Cl^- + As_2O_5,$$

and this reaction enables arsenic to be determined by hypochlorite titration in the same way as the standardisation is carried out.

Ammonia may be determined in bicarbonate medium by hypochlorite titration, the reaction being effectively

$$2NH_3 + 3OBr^- \longrightarrow N_2 + 3Br^- + 2H_2O.$$

The most satisfactory procedure is to add excess of hypochlorite and potassium bromide, and allow the mixture to stand for some time. Sufficient arsenious oxide solution is then added to neutralise the excess hypobromite, and the excess of arsenious oxide is determined by final titration with standard hypochlorite solution.

The iodometric finish may be employed, but the arsenious oxide method has the advantage that foreign oxidising agents such as iron-III, chromate or arsenate do not interfere. On the other hand, iodides interfere with the arsenic method but not with the iodometric method.

Hypochlorite may also be used for the determination of urea, sulphide, thiosulphate, tetrathionate (in alkaline medium), and hydrogen peroxide and nitrite (in weakly alkaline medium). This reaction provides one of the best methods for determining nitrite.

The reactions involved in these determinations are as follows:

$$CO(NH_2)_2 + 3OBr^- \longrightarrow CO_2 + N_2 + 2H_2O + 3Br^-.$$

$$S^- + 4OCl^- \longrightarrow SO_4^- + 4Cl^-.$$

$$S_2O_3^- + 4OCl^- + H_2O \longrightarrow 2SO_4^- + 4Cl^- + 2H^+.$$

$$H_2O_2 + OCl^- \longrightarrow H_2O + O_2 + Cl^-.$$

$$NO_2^- + OCl^- \longrightarrow NO_3^- + Cl^-.$$

The method recommended for these substances is, as in the case of ammonia, to add excess hypochlorite (or hypobromite), and to determine the excess of this in the same way as in the ammonia determination.

Procedures

Standardisation of hypochlorite solution. Prepare an approximately 0·1 N solution of calcium hypochlorite by dissolving 6–10 g.

(depending on the available chlorine content) in 250 ml. water. Filter to remove iron oxide, and other insoluble matter, and dilute to 1 litre. (Alternatively dilute commercially available 10 per cent w/v sodium hypochlorite appropriately.)

Pipette 25 ml. of 0·1 N arsenious oxide solution into a conical flask and add about 1 g. potassium bromide and 0·5 g. sodium bicarbonate. Titrate with the hypochlorite solution until within a few ml. of the anticipated end-point, add one drop 0·2 per cent aqueous solution of Bordeaux B (B.C.I. No. 88) and continue the titration, swirling the flask well between each drop, until the colour changes from pink to colourless or light yellow-green. If the indicator fades before the end-point is reached add one more drop, and continue the titration if any colour remains. Subtract an indicator blank of 0·03 ml. 0·1 N hypochlorite per drop of indicator solution in a final volume of 50–75 ml., or carry out a blank determination.

If iodometric standardisation is preferred, add to 25 ml. hypochlorite solution 1–1·5 g. potassium iodide, and 5 ml. 6 N sulphuric acid, and titrate with standard sodium thiosulphate solution.

Determination of ammonia in ammonium salts. Weigh 1 g. of the salt, dissolve in water in a 250-ml. graduated flask, and make up to the mark. Pipette a 25-ml. portion into a conical flask and add 1 g. potassium bromide and 0·5 g. sodium bicarbonate. Titrate with 0·1 N hypochlorite solution until a permanent light yellow colour due to excess hypochlorite appears. Allow to stand for 5 minutes, and add exactly 10 ml. 0·01 N arsenious oxide solution (prepared by exact dilution of the 0·1 N solution). Add one drop Bordeaux B solution. The colour should remain permanent. If it fades add a further amount of arsenious oxide solution and more indicator. Continue the titration with hypochlorite as in the standardisation against arsenious oxide.

Calculate the amount of hypochlorite solution corresponding to the arsenious oxide solution added, and deduct this volume from the total volume of hypochlorite added. The difference is equivalent to the amount of ammonium present.

$$1 \text{ ml. } 0\cdot1 \text{ N hypochlorite} \equiv 1\cdot135 \text{ mg. NH}_4^+.$$

[1] I. M. Kolthoff and V. A. Stenger, *Ind. Eng. Chem. Anal.*, 1935, **7**, 79.

4. MERCUROUS NITRATE

Mercurous nitrate has been used for many years as a reagent for the titration of chloride and bromide.[1] In 1940 Bradbury and Edwards showed[2] that in the presence of excess of thiocyanate ions mercurous nitrate would reduce iron-III ion and could be used for its direct titration, the end-point being shown by the disappearance of the blood-red colour of complex ferric thiocyanate. Later, Pugh[3] recommended mercurous perchlorate in place of mercurous nitrate, but there seems to be little or no advantage from this substitution.

Although this method for the determination of iron seems to be the simplest yet advanced, the reagent has not come into general use. This is probably because the stoicheiometry of the reaction remained to be proved, and there was no information regarding interferences. Recently, Belcher and West have published their studies on the use of this reagent,[4] and have shown that under selected conditions the reaction, which can be expressed as

$$Fe^{+++}+Hg^+ \longrightarrow Hg^{++}+Fe^{++}$$

is stoicheiometric. They have also investigated the interferences, and in certain cases were able to overcome them. Belcher and West have also shown that the reagent can be used for many other determinations besides that of iron-III. By adding an excess of ferrous ammonium sulphate to certain oxidising agents an equivalent amount of iron-III is produced, and this can be titrated in the usual way. This reaction can be used for standardising the reagent with potassium dichromate. In conjunction with a standard solution of ferric alum, reducing agents such as mercurous mercury and hydroxylamine may be determined. Although in certain of these determinations there is no advantage to be gained over the use of existing reagents, they are described to illustrate the versatility of mercurous nitrate.

The reaction of copper is remarkable, since under the conditions employed it behaves as an oxidising agent to iron-II; normally iron-III oxidises copper-I. This effect is caused by the change in potential due to the excess of thiocyanate ions, and can be predicted theoretically.

The reagent contains too much acid for the titration of halides. However, von Zombory's halide reagent can be used in place of the acid solution of mercurous nitrate, although solutions more concentrated than 0·15 M cannot be prepared. Pugh's mercurous perchlorate reagent can be used for both purposes, but mercurous nitrate is preferred because the reagent is simple to prepare. In any case, mercurometric methods are generally inferior to argentometric methods for the determination of halides.

Nitric acid should not exceed 2·5 N, sulphuric acid 3·8 N, or hydrochloric acid 0·8 N. If, for some reason, it is inconvenient to work below these concentrations, the addition of more ammonium thiocyanate prevents interference. All agents capable of oxidising ferrous iron in the cold, together with copper and molybdenum, interfere. Fluoride, pyrophosphate and oxalate interfere, but their effect can be overcome by increasing the acid concentration. However, if the maximum limits already quoted are overstepped, more ammonium thiocyanate must be added. Nitrites interfere because of the formation of blood-red nitrosyl thiocyanate. This effect can be eliminated by adding an excess of sulphamic acid and warming prior to the addition of ammonium thiocyanate.

For the interference of the less common cations and anions the original papers should be consulted.

Reagents

Mercurous nitrate, 0·1 M. Weigh slightly more than the theoretical amount of mercurous nitrate, dissolve in 200 ml. water containing 50 ml. nitric acid (free from oxides of nitrogen) and dilute to 1000 ml. with dilute nitric acid of the same concentration. The reagent is quite stable.

Ammonium thiocyanate solution. Dissolve 400 g. ammonium thiocyanate in distilled water and dilute to 1000 ml.

Procedures

Determination of iron. Adjust the solution, contained in a 250-ml. conical flask, so that it is approximately 0·5 N in nitric acid. For amounts of iron up to 100 mg. add 10 ml. ammonium thiocyanate solution. For larger amounts add 1 ml. extra for each 10 mg. iron. An excess of ammonium thiocyanate has no ill effect, so that in doubtful cases a considerable excess may be used. Dilute the solution with 0·5 N nitric acid so that the volume at the end of the titration will be approximately 100 ml. Titrate with mercurous nitrate, shaking well until the blood-red turns to orange. At this stage shake well for 15 seconds between addition of each drop, and continue the titration until all trace of ferric thiocyanate has disappeared.

1 ml. 0·1 N mercurous nitrate≡5·584 mg. Fe.

Determination of dichromate, permanganate, vanadate and hydrogen peroxide. Add to the solution an excess of approximately 0·1 N ferrous ammonium sulphate dissolved in 0·1 N sulphuric acid. If this reagent is free from ferric salts the amount added need not be measured accurately; otherwise determine the ferric iron present by a blank determination, and measure accurately the amount added. Add sufficient nitric acid to the solution to make it approximately 0·5 N in that acid and proceed as before. Since the amount of ferric iron produced is equivalent to the amount of oxidising agent originally present, the volume of mercurous nitrate added is also equivalent to the latter.

$$1 \text{ ml. } 0{\cdot}1 \text{ N mercurous nitrate} \equiv 1{\cdot}734 \text{ mg. Cr.}$$
$$\equiv 1{\cdot}099 \text{ mg. Mn.}$$
$$\equiv 5{\cdot}095 \text{ mg. V.}$$
$$\equiv 1{\cdot}701 \text{ mg. } H_2O_2.$$

Determination of copper. Proceed exactly as described for dichromate. A white precipitate of cuprous thiocyanate is present, but this does not obscure the end-point.

1 ml. 0·1 N mercurous nitrate≡6·357 mg. Cu.

Determination of persulphate. Add excess of ferrous ammonium sulphate, warm to about 80° C., cool rapidly and treat as before.

1 ml. 0·1 N mercurous nitrate≡9·606 mg. $S_2O_8^{=}$.

Determination of chlorate. Proceed as in the determination of persulphate, but heat just to boiling.

1 ml. 0·1 N mercurous nitrate≡1·391 mg. ClO_3^{-}.

Determination of mercurous mercury. Add a measured excess of 0·1 N ferric alum in 0·2 N nitric acid (sufficient to give a back titration of 5–10 ml.) to the mercurous solution. Follow this by excess 40 per cent ammonium thiocyanate solution. Titrate as before. Mercuric mercury does not interfere, so that mercurous mercury can be determined in its presence.

1 ml. 0·1 N ferric alum≡20·06 mg. Hg.

Determination of hydroxylamine. Add a 10–30 ml. excess of 0·1 N ferric alum in 1 N sulphuric acid to the solution of hydroxylamine chloride. Wash down the sides of the flask, boil for 3–5 minutes, add 100 ml. 1 N sulphuric acid, and cool the solution. Add 10–15 ml. 40 per cent ammonium thiocyanate solution and titrate as before.

$$1 \text{ ml. } 0\text{·}1 \text{ N ferric alum} \equiv 1\text{·}652 \text{ mg. NH}_2\text{OH.}$$

Standardisation of mercurous nitrate. Proceed as in the determination of dichromate, using a standard solution of potassium dichromate.

[1] R. Burstein, *Z. anorg. Chem.*, 1928, **168**, 325: L. von Zombory, *ibid.*, 1929, **184**, 237.
[2] F. R. Bradbury and E. C. Edwards, *J. Soc. Chem. Ind.*, 1940, **59**, 96 T.
[3] W. Pugh, *J.C.S.*, 1945, 588.
[4] R. Belcher and T. S. West, *Anal. Chim. Acta*, 1951, **5**, 260, 268, 360, 364, 472, 474, 546: 1952, **7**, 470.

5. DIPHENYLCARBAZONE COMPLEXES

When potassium ferrocyanide is added to a zinc solution in the ordinary method of titration, using an external indicator and a pH of 6·0 or less, 1·5 atoms of zinc are precipitated by a molecule of ferrocyanide. If the pH is accurately controlled at pH 8·0, 2 atoms of zinc are precipitated by each molecule of ferrocyanide. Under these conditions zinc forms a red complex with diphenylcarbazone which is extractable by organic solvents and whose colour is discharged on reaching the end-point of a ferrocyanide titration.[1]

The reagent is standard potassium ferrocyanide, which is quite stable if kept in an amber glass bottle, and which may be standardised against a standard zinc solution. Cadmium, cobalt, copper, iron-II and nickel interfere. The interference of small traces of nickel, which is very common, is eliminated by addition of cyanide, which complexes both nickel and zinc, followed by acetone which breaks down the zinc complex without affecting the nickelocyanide.

For determination of amounts of nickel below 1 mg. the ordinary cyanide titration is not sufficiently sensitive, and a colorimetric determination is time-consuming and requires considerable care to obtain satisfactory accuracy. Using the red colour formed by nickel with diphenylcarbazone, nickel down to amounts of a few micrograms may be determined titrimetrically by cyanide titration.

The following interfere and must be absent: cadmium, cobalt, copper, iron-II, mercurous, mercuric, vanadium-IV, vanadium-V and zinc.

Procedures

Determination of zinc. Rinse two 250-ml. stoppered Pyrex flasks with dilute sodium hydroxide solution and then with distilled water. Neutralise the sample solution so that it gives the faintest mauve with litmus paper, and add it to one of the flasks. To each of the flasks add 20 ml. 20 per cent ammonium nitrate solution, 15 ml. 0·2 N sodium carbonate solution, 4 drops 10 per cent potassium cyanide solution and sufficient water to bring the bulk of solution in each flask up to 150 ml. Add 10 ml. acetone to each flask.

Shake both mixtures and allow to stand for 15 minutes. Add to each flask 10 ml. amyl alcohol, 10 ml. carbon tetrachloride and 0·3 ml. 1·5 per cent ethanolic solution of diphenylcarbazone. For very small amounts of zinc these last three additions may be halved. Shake vigorously for 15 seconds. Add in small portions from a burette to the flask containing the sample 0·01 M potassium ferrocyanide (or 0·001 M for amounts of zinc less than 2 mg.). Shake vigorously between each addition. After shaking invert both flasks, allow the liquid to settle and become free from bubbles, and compare the colours of the two solvent layers in the necks of the flasks. Take as the end-point the point where the red colour has completely disappeared and the shade in the flask containing the sample is equal to or less than that in the comparison flask.

The solution in the comparison flask fades after some hours, and should not, in any case, be used for more than a day.

1 ml. 0·01 M potassium ferrocyanide ≡ 1·3076 mg. Zn.

Determination of nickel. Wash two stoppered Pyrex flasks, as described for the zinc determination. In one, place the neutralised sample containing 1 mg. nickel or more, and to each flask add 20 ml. 20 per cent ammonium nitrate solution, 15 ml. 0·2 N sodium carbonate solution, and sufficient water to bring the bulk of the solution in each flask to 100–150 ml. Add 20 ml. amyl alcohol, and 0·3 ml. 1·5 per cent ethanolic diphenylcarbazone solution to each flask. Shake vigorously.

Titrate the flask containing the sample with a standard solution containing 4·8 g. potassium cyanide and 2·3 g. potassium hydroxide per litre, following the procedure described for the titration of zinc.

Standardise the potassium cyanide solution frequently against silver nitrate solution, using potassium iodide as indicator.

1 ml. potassium cyanide solution ≡ 1·00 mg. Ni.

For amounts of nickel between 1 mg. and 0·1 mg. use a ten-fold dilution of the standard cyanide solution. For amounts below 0·1 mg. use a 100-fold dilution of the cyanide solution, and add 50 ml. sodium carbonate solution instead of 15 ml. In these low concentrations observe the colours during titration against a white-tile background.

[1] B. S. Evans, *Analyst*, 1946, **71**, 455.

6. STANNOUS CHLORIDE

The reaction of stannous chloride solutions with ferric ions has for long been used as a titrimetric reaction for the determination of iron-III. Szabó and Sugár[1] describe in detail a method for the determination of the end-point, based on the fact that when most of the iron has been reduced the addition of ammonium molybdate in the presence of phosphate results in the formation, at the end-point, of molybdenum blue produced by the excess of stannous chloride.

The phosphate ion increases the oxidation potential of the molybdate, so that practically all the iron-III ions must first be removed to prevent precipitation of the phosphate as a complex. To achieve this, thiocyanate is used as a preliminary indicator.

The standard stannous chloride must be prepared and stored in an oxygen-free atmosphere. Special apparatus is used for this purpose.

The indicator reaction is reversible. The optimum temperature for the reaction is 60°–70° C., and the acid concentration should correspond to a pH of about 0·6. The presence of ammonium chloride speeds up the reaction.

Aluminium, antimony-V, arsenic, lead, manganese, tungsten and zinc do not interfere, nor do small amounts of copper; but copper in amounts greater than iron produces an uncertain end-point. The reaction cannot be used in the presence of vanadium.

This method is recommended for the determination of iron in iron ores.

The use of standard stannous solution has been extended to the titrimetric determination of a number of other ions,[2] including bromate, dichromate, ferricyanide, iodate and vanadate. It may also be used for the determination of iodine.

Solutions required

Standard stannous chloride. Place 80 ml. concentrated hydrochloric acid in a 1-litre storage bottle connected directly to a burette, and add several pieces of marble to expel oxygen. Add 12 g. crystal-

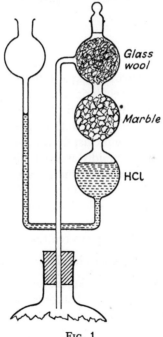

Glass wool

Marble

HCL

Fig. 1

line stannous chloride, and when it has dissolved make the solution up to 1 litre. Maintain a carbon dioxide atmosphere over the solution by fitting to the stopper the apparatus shown in Fig. 1, which acts as a small carbon dioxide generator when liquid is removed from the storage bottle to the burette.

To standardise the solution, place 1 ml. concentrated hydrochloric acid and several small pieces of marble in a narrow-necked titration flask, add one or two crystals of solid potassium bromide and four drops rubrophen or methyl orange indicator solution, and run in 20 ml. standard stannous chloride solution. Titrate rapidly with 0·1 N potassium bromate solution until the solution in the titration flask is colourless.

If properly stored, the stannous solution is stable for several months.

Potassium thiocyanate. Dissolve 4·85 g. in 100 ml. water.
Ammonium molybdate. Dissolve 1·96 g. in 100 ml. water.
Sodium phosphate. Dissolve 1·1 g. in 100 ml. water.
Ammonium phosphomolybdate. Dissolve 1·9 g. in 100 ml. dilute ammonia solution.

Procedures

Determination of iron. Place 20 ml. ferric solution containing 3–150 mg. iron in a narrow-necked 100-ml. titration flask. Add 1–2 ml. concentrated hydrochloric acid and 1 g. solid ammonium chloride. Heat to 60°–70° C., add 2–3 drops of a solution of potassium thiocyanate containing 4·85 g. per 100 ml., and several pieces of marble. Titrate immediately with standard stannous solution while the gas is being rapidly evolved, until the colour is straw yellow. Add 4 drops ammonium molybdate solution and 3 drops sodium phosphate solution (or, instead of these two solutions, 3 drops of an ammonium phosphomolybdate solution) and continue titration until the colour changes from green to blue. Add the last few drops slowly to allow complete reaction.

1 ml. 0·1 N stannous chloride≡5·584 mg. Fe.

Determination of dichromate. Adjust the volume of dichromate solution, containing 50–150 mg. dichromate, to 20 ml. Add 8 ml. concentrated hydrochloric acid, 3 drops diphenylamine indicator solution in concentrated sulphuric acid, and a few fragments of marble. Titrate to a green colour.

1 ml. 0·1 N stannous chloride≡4·9035 mg. $K_2Cr_2O_7$.

Determination of vanadate. To 20 ml. of the solution, containing

50–250 mg. vanadate, add 10 ml. concentrated hydrochloric acid and 3 drops diphenylamine indicator solution. Add sufficient marble (about 2 g.) to maintain vigorous effervescence throughout the titration. Titrate to the point where the violet solution changes to light green.

1 ml. 0·1 N stannous chloride≡11·699 mg. NH₄VO₃.

Determination of iodate. Adjust the volume of iodate solution, containing 30–100 mg. iodate, to 20 ml. Add 2 ml. concentrated hydrochloric acid, and titrate immediately with stannous chloride solution to pale yellow. Add starch indicator and complete the titration to the first decolorisation of the starch-iodide complex.

1 ml. 0·1 N stannous chloride≡3·5669 mg. KIO₃.

Determination of bromate. (*a*) Adjust the volume of bromate solution, containing 20–100 mg. bromate, to 20 ml. Add 2 ml. concentrated hydrochloric acid and titrate immediately with stannous chloride solution till the bromine colour disappears.

(*b*) After the addition of the hydrochloric acid add 0·2 g. potassium iodide and complete as for iodate. This method is more convenient, but less reproducible than (*a*).

1 ml. 0·1 N stannous chloride solution≡2·7835 mg. KBrO₃.

Determination of ferricyanide. Adjust the volume of the ferricyanide solution, containing at least 250 mg. ferricyanide, to 20 ml. Add 8 ml. concentrated hydrochloric acid and a few pieces of marble. Titrate rapidly to a green colour, and then add the stannous chloride solution dropwise till a clear blue is obtained.

1 ml. 0·1 N tannous chloride≡32·9236 mg. K₃Fe(CN)₆.

Determination f iodine. Adjust the volume of iodine solution, containing 10–250 mg. iodine, to 20 ml. Add 3 ml. concentrated hydrochloric acid, and titrate with stannous chloride solution to pale yellow. Add starch indicator solution and complete the titration to the first decolorisation of the starch-iodide complex.

1 ml. 0·1 N stannous chloride≡12·692 mg. I₂.

[1] Z. Szabó and E. Sugár, *Analyt. Chem.*, 1950, **22**, 361.
[2] *Idem, Anal. Chim. Acta*, 1952, **6**, 293.

7. POTASSIUM PERIODATOPERCUPRATE

It has been known for many years that copper can form certain complex compounds by utilising a valency higher than that normally associated with the simple ions. The history up to 1925 of compounds containing tervalent copper has been reviewed by Votis.[1] Malaprade[2] succeeded in isolating the sparingly soluble sodium diperiodatocuprate with a formula of $Na_7[Cu(IO_3)_2]$. From the work of Votis and Malaprade, and from further work of Malatesta[3] it may be concluded that copper may be oxidised from the bivalent to the tervalent state by means of persulphate, but that the tervalent ion is unstable, and requires to be stabilised by co-ordination with a suitable anion. This stabilising function may be performed by periodate or tellurate, the former being particularly suitable owing to its greater stability and ease of preparation.

A number of interesting determinations are possible using the percuprate reagent.

Naphthalhydroxamic acid forms precipitates with a number of cations, but, with the exception of the calcium, strontium and barium compounds, these can all be dissolved in ammonium hydroxide, ammonium tartrate or ammonium nitrilotriacetate. The naphthalhydroxamate of barium is yellow; those of calcium and strontium form first as yellow precipitates but rapidly turn brick-red. The calcium compound is so sparingly soluble that unlike the oxalate it can be precipitated from solutions containing tartrate, citrate or nitrilotriacetate ions, or from hard water which has been diluted to fifty times its volume. It can even be prepared by boiling such "insoluble" calcium salts as the oxalate, phosphate, carbonate or fluoride with a solution containing the naphthalhydroxamate of an alkali metal. The concentration limit for the precipitation of simple calcium ions is 0.1 μg./ml.

The use of naphthalhydroxamates for the gravimetric or colorimetric determination of calcium has been described by Beck,[4] and presents no unusual features; but the adaptation of the method to a titrimetric procedure, which appeared desirable, presented a considerable amount of difficulty. In spite of its low solubility, calcium naphthalhydroxamate is readily decomposed with the liberation of the free acid; the stability of this acid, as well of its salts, towards oxidising agents, however, renders its titrimetric determination no easy matter. Most of the investigations have centred on the direct titration of the calcium naphthalhydroxamate, but permanganate, dichromate, iodine, or ferric salts failed to give good results. The substance is slowly oxidised by sodium hypochlorite, but the final iodometric titrations were inconsistent. The only successful oxidant was potassium diperiodatocuprate in alkaline solution. The oxidation with this reagent takes place in two stages: the first, involving four atoms of copper per molecule of calcium naphthalhydroxamate, takes place quite rapidly, and results in conversion to naphthalic acid and nitrite ions; the second stage takes place very slowly, involves two atoms of copper, and results in oxidation of nitrite to nitrate. The latter reaction takes place more rapidly at pH 6·0–7·0.

Several other inorganic substances have been determined by direct titration with potassium periodatocuprate. These include some of the common reducing agents, such as arsenious oxide, antimonious oxide and sodium thiosulphate, as well as the sulphides of such elements as arsenic, antimony, phosphorus, molybdenum, tin, lead and zinc, in the form of alkaline solutions of the (higher) sulphides. The reaction suggested for the oxidation of thiosulphate is

$$S_2O_3^{=}+2OH^{-}+4O \longrightarrow 2SO_4^{=}+H_2O,$$

while that for the titration of arsenic sulphide appears to be

$$As_2S_3+2O+6OH^{-} \longrightarrow AsS_2O_2^{=}+AsSO_3^{=}+3H_2O.$$

In boiling solution a further oxidation takes place more slowly represented by

$$AsS_2O_2^{=}+4O+H_2O \longrightarrow AsSO_3^{=}+H_2SO_4.$$

Of greater significance is the titration of cyanides, the procedure affording a method for determining cyanides without interference from halides. The reaction appears to be

$$CN^{-}+2OH^{-}+5O \longrightarrow CO_3^{=}+NO_3^{-}+H_2O.$$

Chromic salts can be titrated with potassium diperiodatocuprate, but although the oxidation proceeds rapidly and smoothly the end-point is difficult to detect owing to the intense yellow colour of the chromate formed in the reaction. For this titration, therefore, potassium periodatonickelate is recommended in preference to the copper compound.

14

The reagent may be standardised by titration against glucose solution. One molecule of glucose reacts with eight atoms of copper. In addition to being employed for determination of glucose, the reagent may be employed for titrimetric determination of other organic compounds such as reducing sugars, amino-acids and proteins, in such biological fluids as blood or urine.[5]

Procedures

Preparation of the reagent.[5] Dissolve two parts potassium periodate and one part copper sulphate pentahydrate in water, and add ten parts potassium hydroxide. Boil the mixture with one-half part potassium persulphate until the colour becomes dark brown to black. Remove insoluble matter by centrifuging the solution, and standardise against glucose solution. Potassium salts must be used throughout. Sodium compounds are unsuitable since they give rise to a precipitate of sodium diperiodatocuprate.

Potassium iodate may be used instead of periodate provided that an appropriately larger amount of persulphate is used. Similarly, in the preparation of the tellurato-compound the more readily available tellurium dioxide or potassium tellurite may be substituted for potassium tellurate. These substances are dissolved in potassium hydroxide, and after addition of copper sulphate the mixture is oxidised with potassium persulphate to give ditelluratocuprate $K_5H_4[Cu(TeO_6)_2]$.[3] Once prepared, these solutions are stable for several days.

Standardisation of the solution.[5] Use a solution containing 0·1 per cent glucose. Take an aliquot (0·2–0·3 ml.) and add 30 ml. water and three pellets of pure potassium hydroxide. Titrate, using a microburette, to a green colour which is stable for 1 minute. The titration can be carried out in the cold, and atmospheric oxygen does not interfere.

Determination of calcium.[4, 6] Boil an approximately 0·5 N solution of calcium for a short time with sodium naphthalhydroxamate. Separate the precipitate by centrifuging. Wash successively with 20 per cent, 50 per cent and pure acetone, and suspend in 50 per cent potassium hydroxide solution. Titrate the warm solution with 0·01 N reagent. The end-point is very sharp, being marked by the appearance of a green colour arising from the combination of the colours of bivalent and tervalent copper. It is detectable with an excess of 0·005 ml. titrant, corresponding to 0·1 μg. calcium.

Determination of arsenic, antimony, thiosulphate, sulphide or cyanide.[4] Measure out 0·5 ml. solution and add 5 ml. 10 per cent potassium hydroxide solution. Titrate with the reagent to the green end-point, using a microburette. In the case of the cyanide a transient violet colour appears at the beginning of the titration.

[1] M. Votis, *Rec. Trav. chim.*, 1925, **44**, 425.
[2] L. Malaprade, *Compt. rend.*, 1937, **204**, 979.
[3] L. Malatesta, *Gazzetta*, 1941, **71**, 467, 580.
[4] G. Beck, *Mikrochem. Mikrochim. Acta*, 1951, **36–37**, 245.
[5] *Idem, ibid.*, 1950, **35**, 169.
[6] *Idem, Anal. Chim. Acta*, 1950, **4**, 245.

8. ETHYLENEDIAMINETETRA-ACETIC ACID

As indicated elsewhere (pp. 93 ff.) ethylenediaminetetra-acetic acid promises to be a most valuable reagent for analytical purposes, and indeed, probably to be the most versatile titrimetric reagent known.

It may be that many of the methods in which it is used require further study before they can be applied with complete confidence. Therefore the titrimetric applications described here in detail will be confined to the determination of water hardness, which has been extensively investigated.

Ethylenediaminetetra-acetic acid forms stable colourless complexes with calcium and magnesium.[1] In addition, murexide (ammonium purpureate), which has a blue-violet colour in alkaline solution, becomes salmon-pink in the presence of calcium ions. When the calcium is complexed with EDTA the original colour of the murexide is restored. Accordingly calcium may be titrated with EDTA using murexide as indicator. Several other metal ions, *e.g.*, cadmium, copper, mercury and zinc, give a colour change with murexide, but magnesium is without effect. On this basis a method has been devised for the determination of the calcium hardness of waters in which the interfering ions would normally be absent. The titration is effected at a pH of about 12·0. This is attained by the addition of sodium hydroxide solution. Magnesium, if present, is in the non-ionic form and may be precipitated.

An aqueous solution of murexide is unstable because of oxidation. Betz and Noll[2] tried various means of overcoming this drawback and eventually recommended that the solid murexide be diluted with sodium chloride and the solid mixture added to the solution immediately before the titration.

Biedermann and Schwarzenbach[3] showed that Eriochrome Black T could be used as indicator in the titration of the alkaline earth metals and certain other metals. At a pH of 8·0–10·0 the blue colour of the dye changes to wine-red in the presence of calcium or magnesium. The complex formed with calcium is too unstable for indicator purposes, but that with magnesium is much more stable and gives satisfactory end-points. When a solution containing calcium and magnesium is titrated with EDTA in the presence of Eriochrome Black T, the calcium is preferentially complexed, but an end-point is not obtained until all the magnesium has also been complexed.

Eriochrome Black T is the same substance as Solochrome Black WDFA (B.C.I. No. 203). In the determination of total hardness it is used at a pH of approximately 10·0. This pH is obtained by the use of a buffer. Schwarzenbach[1] used an ammonium hydroxide-ammonium chloride buffer, and Diehl, Goetz and Hach[4] recommend a similar one. Betz and Noll,[2] however, prefer a sodium tetraborate-sodium hydroxide buffer. Diehl, Goetz and Hach dissolved the indicator in ethanol, whereas Betz and Noll use a diluted *iso*-propanol solution. Schwarzenbach dissolved the indicator in the buffer solution, but found it to be unstable.

Magnesium hardness may be determined after precipitating calcium with a soluble oxalate. The insoluble calcium oxalate may be filtered off, or the titration can be carried out in its presence. In the latter case it is necessary to titrate rapidly since EDTA dissolves calcium oxalate.

Small amounts of metals other than calcium and magnesium may be present in natural waters, and iron, aluminium, copper and manganese may interfere. As little as 0·1 p.p.m. manganese causes fading of the Solochrome Black WDFA indicator, and copper prevents the change of colour. Various suppressing agents have been proposed to overcome these interferences, sodium sulphide and sodium cyanide being two of these. Heald, Coates and Edwards[5] examined the various modifications for the determination of hardness using EDTA, and included in their study an examination of the interfering effect of various ions. They concluded that the addition of suppressing agents was unnecessary for the quantities normally encountered.

The effect of interfering substances is summarised in Table VIII.

TABLE VIII

Interfering ion	Total hardness determination	Calcium hardness determination
Cu^{++}	No end-point when more than 0·2 p.p.m.	Cu^{++} titrated as hardness but otherwise no interference
Fe^{+++}	Good results in presence of 20 p.p.m. End-point became greyer as content increased to this value.	Reasonable results in presence of 5 p.p.m. At higher concentrations low results were obtained, possibly due to co-precipitation of Ca^{++} with Fe^{+++}.
Al^{+++}	End-point became redder as content increased, but good results obtained up to 50 p.p.m.	No interference up to 50 p.p.m.
Mn^{++}	Mn^{++} titrated as hardness, but caused no interference otherwise up to 2 p.p.m.	Mn^{++} was oxidised and precipitated when 2 p.p.m. were present, but no interference was experienced.
MnO_4^-	Hydroxylamine (added to the indicator) reduced MnO_4^- to Mn^{++} which titrated as hardness.	2 p.p.m. masked the colour of the end-point, and titration was not possible.
PO_4^-	Amounts up to 50 p.p.m. (the highest amount tested) caused no interference.	Low results in the presence of 10 p.p.m. With higher concentrations very low results were obtained due to precipitation of calcium phosphate.
Polyphosphate (Calgon)	Reasonable results up to 10 p.p.m. With higher concentrations the results were low.	Reasonable results up to 10 p.p.m. With higher concentrations the results were low.

Reagents

Titrating solution. 0·02 N Disodium ethylenediaminetetraacetate dihydrate: Dissolve 3·72 g. crystalline EDTA hydrate in distilled water and dilute to 1 litre. Check the solution against calcium chloride solution which has been standardised gravimetrically.

Ammonia buffer solution. Dissolve 67·5 g. ammonium chloride in 570 ml. ammonia solution (sp. gr. 0·88) and dilute to 950 ml. with distilled water. To this add a solution of 0·616 g. magnesium sulphate heptahydrate and 0·93 g. solid EDTA reagent in 50 ml. water. The use of 2 ml. of this solution in 100 ml. water is equivalent to the addition of 5 p.p.m. of magnesium in terms of $CaCO_3$ together with its equivalent of EDTA.

Sodium hydroxide solution. Approximately 4 N.

Total hardness indicator. Dissolve 0·5 g. Solochrome Black WDFA and 4·5 g. hydroxylamine hydrochloride in 100 ml. ethanol.

Calcium hardness indicator. Mix 0·20 g. murexide and 100 g. sodium chloride and grind to a fine powder.

Bromophenol blue indicator solution (p. 158).

Hydrochloric acid solution. 25 per cent w/w.

Activated carbon. Activated Charcoal for Decolorising Purposes (B.D.H.). A blank determination should give less than 0·5 p.p.m. If it does not the sample should not be used.

Procedures

(*i*) *For raw and untreated waters of hardness greater than 10 p.p.m. in terms of* $CaCO_3$. *Total hardness.* Transfer 100 ml. of the sample to a porcelain dish, add 2 ml. ammonia buffer solution and 6 drops of Solochrome Black solution, and titrate until all trace of red colour disappears. The final colour at the end-point is usually pure blue, but with some waters a neutral grey end-point is obtained. When the hardness of the water is above 250 p.p.m. $CaCO_3$, take a 50-ml. portion for the titration. The sample is not diluted with distilled water unless some interference is experienced with the end-point.

$$\text{Total hardness (p.p.m. } CaCO_3) \equiv \frac{\text{ml. titrating solution} \times 1000}{\text{ml. sample}}.$$

Calcium hardness. Transfer 100 ml. sample to a porcelain dish, add 1 ml. sodium hydroxide solution, and follow this with 0·2 g. murexide indicator added by means of a graduated scoop. Titrate until the colour of the solution changes to violet. The end-point is best indicated when the addition of a further 0·1 ml. titrating solution produces no further change in colour. When the calcium hardness exceeds 250 p.p.m. $CaCO_3$, take a 50-ml. portion for the titration. The sample is not diluted with distilled water unless some interference is experienced at the end-point.

$$\text{Calcium hardness (p.p.m. } CaCO_3) \equiv \frac{\text{ml. titrating solution} \times 1000}{\text{ml. sample}}.$$

(*ii*) *For raw and untreated waters of hardness less then 10 p.p.m. in terms of* $CaCO_3$. *Total hardness*. Add 5 ml. of "standard hard water" (or about 50 ml. of a natural water) to 50 ml. distilled water contained in a large porcelain dish. Add 10 ml. ammonia buffer and 6 drops of Solochrome Black solution, and titrate precisely to the end-point. To the solution add 500 ml. of the low-hardness water and a further 1 ml. indicator, and continue the titration to the end-point. The difference between the titration figures is a measure of the total hardness.

$$\text{Total hardness (p.p.m. } CaCO_3) \equiv \frac{\text{ml. difference in titres} \times 1000}{\text{ml. sample}}.$$

Calcium hardness. Add 5 ml. sodium hydroxide solution and 0·6 g. murexide indicator to 500 ml. of the low-hardness water contained in a large porcelain dish. Titrate as described in the previous calcium-hardness determination.

$$\text{Calcium hardness (p.p.m. } CaCO_3) \equiv \frac{\text{ml. titrating solution} \times 1000}{\text{ml. sample}}.$$

Magnesium hardness. In both (*i*) and (*ii*) magnesium hardness is obtained by difference.

(*iii*) *For boiler water of low colour*. *Total hardness*. Use method described above under (*ii*).
Calcium hardness. Use method described above under (*ii*).
Magnesium hardness. This is obtained by difference, but as magnesium hardness is generally absent the total hardness determination is usually sufficient.

(*iv*) *For boiler waters of high colour*. Add 1 g. activated carbon to 100 ml. distilled water containing 1 ml. hydrochloric acid solution.

Mix well and filter on a paper-pulp pad supported on a Witt filter-disc. Filter under moderate vacuum and wash with 100 ml. distilled water. Filter 500 ml. of the coloured boiler water and adjust with hydrochloric acid solution to a pH of about 2·5, *i.e.*, until a small test portion gives a yellow colour with bromophenol blue indicator. Filter the water through the bed of charcoal, refiltering if necessary until the colour is satisfactory. Determine the total and calcium hardness as in (*ii*) above, using double the normal amount of the ammonia buffer and sodium hydroxide solution to allow for the acidity of the water.

[1] G. Schwarzenbach, W. Biedermann and F. Bangerter, *Helv. Chim. Acta*, 1946, **29**, 811.

[2] J. D. Betz and C. A. Noll, *J. Amer. Water Works Assoc.*, 1950, **42**, 49.

[3] W. Biedermann and G. Schwarzenbach, *Chimia*, 1948, **2**, 56.

[4] H. Diehl, C. A. Goetz and C. C. Hach, *J. Amer. Water Works Assoc.*, 1950, **42**, 40.

[5] I. A. Heald, K. B. Coates and J. E. Edwards, *Ind. Chem. Chem. Mfr.*, 1950, **26**, 428.

CHAPTER VII

MISCELLANEOUS METHODS

1. TITRIMETRIC DETERMINATION OF BROMIDE

KOLTHOFF AND YUTZY[1] have described a method for the determination of bromide in the presence of much chloride which is based on the earlier work of van der Meulen[2] and of D'Ans and Höfer.[3] The method consists of oxidising the bromide to bromate with sodium hypochlorite in the presence of a sodium dihydrogen phosphate buffer, destroying the excess hypochlorite with sodium formate, and determining the bromate iodometrically after adding ammonium molybdate as catalyst to accelerate the liberation of the iodine. The method is very sensitive because of the six-fold amplification of the normal titration figure.

In the original method of van der Meulen a large excess of sodium chloride was added. Kolthoff and Yutzy found this to be unnecessary when the amount of bromide was 4 mg. or less, and indeed to be undesirable because of the presence of bromide impurity in the purest sodium chloride obtainable. For amounts of bromide greater than 4 mg. the addition of sodium chloride appears to be necessary since otherwise low results are obtained.

Willard and Heyn[4] have applied the method to the determination of bromine in brines containing much calcium and magnesium. To avoid precipitation of phosphate they used an acetic acid-sodium acetate buffer, and established that the pH range necessary for quantitative oxidation of bromate is pH 5·5–7·0. In an actual determination the pH is adjusted either using a glass electrode or by adding a zinc salt and then acetic acid until the precipitate is just dissolved. The solution is acidified with hydrochloric acid instead of sulphuric acid, prior to the titration, to avoid precipitation of calcium sulphate.

An extensive examination of the Kolthoff-Yutzy procedure has been made by Haslam and Moses[5] who wished to apply the method to the analysis of brines. They found that the average recovery was 99·5 per cent in the presence of 15 g. sodium chloride. Satisfactory results were obtained for a mixture containing 0·246 g. CaO and 0·10 g. MgO. For a mixture containing 0·492 g. CaO and 0·20 g. MgO low results were obtained. This is attributed to the difficulty of obtaining complete oxidation in the presence of heavy precipitates of calcium and magnesium phosphates. Presumably the Willard-Heyn procedure would be preferable under these conditions. The effect of the pH on the oxidation was re-examined. The latter was quantitative over the range pH 4·12–7·35. When 0·2 g. MgO (as the sulphate) and 0·492 g. CaO (as the chloride) were present the pH values of the solutions were 4·52 and 4·49 respectively. In the absence of both the pH was 5·4. The upper limit of potassium bromide which was quantitatively oxidised by 10 ml. N sodium hypochlorite in the absence of sodium chloride was 31·7 mg. (21·3 mg. Br). This is proportionately greater than the amount found by Kolthoff and Yutzy, even allowing for the larger amount of oxidant used. In the presence of 15 g. sodium chloride the amount was 79·3 mg. KBr (53·3 mg. Br).

Alicino, Crickenburger and Reynolds[6] have applied the Kolthoff-Yutzy procedure to the micro-determination of bromine in organic compounds after decomposition in a combustion tube and absorption in sodium hydroxide. They found that satisfactory results could also be obtained when decomposition was effected in the Parr bomb. White and Kilpatrick[7] decompose the sample in a Carius tube and neutralise the acid digestion mixture prior to oxidation, but otherwise follow the same procedure. Since iodide is oxidised to iodate by this procedure, the final value would, in the presence of iodine, correspond to bromine plus iodine. An assessment of both could probably be made by a separate determination of iodine, using Leipert's method, which is selective for iodides.

In the Kolthoff-Yutzy and Willard-Heyn procedures, sodium hypochlorite is prepared from chlorine and sodium hydroxide. A commercial sodium hypochlorite ("Clorox" for domestic purposes, containing 5·25 per cent NaClO) was used by Alicino and his coworkers, and by White and Kilpatrick. Haslam and Moses used a commercial bromine-free product (I.C.I. General Chemicals, Widnes) containing 14–15 per cent NaClO. They determined the actual hypochlorite content by titration against standard arsenious oxide solution; and the sodium hydroxide content by adding 30 ml. neutral 6 per cent hydrogen peroxide and 5 ml. neutral 0·5 M barium chloride to 10 ml. of the original solution to destroy hypochlorite, and then titrating with 0·1 N hydrochloric acid using phenolphthalein as indicator. The solution was then diluted to make it N in hypochlorite and 0·1 N in sodium hydroxide.

Procedures

The Kolthoff-Yutzy Modification. Adjust the neutral solution to a bulk of 25 ml. or less. Add 1 g. sodium dihydrogen phosphate, 10 g. sodium chloride and 5 ml. N sodium hypochlorite. Heat the solution just to boiling and add 5 ml. 50 per cent sodium formate. Cool, dilute to 150 ml., add 1 g. potassium iodide, 25 ml. 6 N sulphuric acid and 1 drop 0·5 N ammonium molybdate solution, and titrate immediately with standard sodium thiosulphate solution.

Run a blank under the same conditions and deduct the value obtained from that of the actual determination.

The sodium chloride may be omitted when the bromine is less than 2 mg.

1 ml. 0·01 N sodium thiosulphate ≡ 0·1332 mg. Br.

The Haslam-Moses Modification. Neutralise 50 ml. solution to methyl red. Add 2·0 g. sodium dihydrogen phosphate dihydrate and 10 ml. N sodium hypochlorite solution, and bring just to the boil. Remove from the source of heat, and add 10 ml. sodium formate solution (32 g. sodium hydroxide, prepared from sodium metal, dissolved in 32 ml. 90 per cent formic acid with ice cooling, and made up to 100 ml. with water). Wash down the sides of the beaker and the cover-glass, and allow to stand for 5 minutes. Cool, and add 240 ml. water, 2·0 g. potassium iodide, 50 ml. 6 N sulphuric acid and 1 drop ammonium molybdate solution (2·9 g. dissolved in water and made up to 100 ml. with water). Titrate with 0·1 N sodium thiosulphate solution.

Presumably the 15 g. sodium chloride is added only when the potassium bromide is more than 31·7 mg.

The Alicino-Crickenburger-Reynolds Modification. Use sodium hydroxide solution to absorb the products of combustion. The pH will not vary from 6·2 for 0–2 ml. N sodium hydroxide. Sodium chloride is not added, being unnecessary for the amounts of bromine involved. Treat the solution and washings from the absorption tube (20–30 ml.) with 5 ml. of a solution containing 20 g. sodium dihydrogen phosphate dihydrate in 100 ml. water and with 5 ml. Clorox. Heat just to boiling. Add 5 ml. 50 per cent sodium formate, cool the solution, and add 10 ml. 9 N sulphuric acid, 1·0 g. potassium iodide and 1 drop 0·5 N ammonium molybdate solution. Titrate the solution with standard thiosulphate solution.

The Willard-Heyn Modification. Dilute a sample containing not more than 13 mg. bromine to 25–30 ml. in a 400-ml. beaker. If the solution is not neutral add sodium hydroxide or acetic acid. Add 5 g. sodium chloride. Depending on the amount of bromine present add 5–10 ml. N sodium hypochlorite. Using the glass electrode adjust the pH to 6·0–6·5 by adding sodium hydroxide solution or glacial acetic acid. Alternatively add 0·1 g. zinc acetate dissolved in 5 ml. water containing a few drops acetic acid, and then add glacial acid so that the precipitate of zinc hydroxide just dissolves. Rinse the sides of the beaker and bring to the boil in 5–10 minutes. Add 5 ml. 50 per cent sodium formate solution. The solution should effervesce indicating that excess hypochlorite was present. Rinse the beaker sides and boil for a few seconds. Cool, dilute to 150–200 ml. with water, and add 5 ml. 20 per cent potassium iodide solution or 1·0 g. solid potassium iodide. If free iodine is liberated at this stage discard the test. Add 5 ml. concentrated hydrochloric acid diluted to 30 ml. with water, and 1–2 drops 0·25 M ammonium molybdate solution. Titrate immediately with standard sodium thiosulphate solution.

Run a blank and deduct the value in the usual way.

[1] I. M. Kolthoff and H. Yutzy, *Ind. Eng. Chem. Anal.*, 1937, **9**, 75.
[2] J. H. van der Meulen, *Chem. Weekblad*, 1931, **28**, 82, 238: 1934, **31**, 558.
[3] J. D'Ans and P. Höfer, *Z. angew. Chem.*, 1934, **47**, 73.
[4] H. H. Willard and A. H. A. Heyn, *Ind. Eng. Chem. Anal.*, 1943, **15**, 321.
[5] J. Haslam and G. Moses, *Analyst*, 1950, **75**, 343.
[6] J. F. Alicino, A. Crickenburger and B. Reynolds, *Analyt. Chem.*, 1949, **21**, 755.
[7] L. M. White and M. D. Kilpatrick, *ibid.*, 1950, **22**, 1049.

2. TITRIMETRIC DETERMINATION OF CADMIUM

The determination of cadmium in the presence of zinc can be achieved rapidly and accurately by precipitation with brucine sulphate and potassium iodide.[1] The iodide in the precipitate is then titrated with silver nitrate, using eosin Y as indicator. The method may be used for the determination of cadmium in sulphide ores after removal of copper, iron and lead, which must be absent. Nitrate must also be absent. Zinc in a ratio as high as 1000 : 1 does not interfere.

In making up the brucine sulphate solution it is advisable to add up to 5 per cent sulphuric acid. For the best results the solution should be prepared fresh each day.

A large bulk of precipitate may retain some of the iodide solution, so that the amount of cadmium present should not be too great.

Procedure

Determination of cadmium. Adjust the neutral or slightly acid solution to contain 20–50 mg. cadmium in a total bulk of 50 ml. Add 1·5 ml. 1 per cent brucine sulphate solution for each mg. cadmium believed to be present, and then add 1·5 ml. 10 per cent potassium iodide solution. Stir well. Allow to stand for 10 minutes, and filter through a Büchner funnel using a fine paper. Wash three times with a 1 : 1 mixture of the brucine sulphate and potassium iodide solutions (prepared immediately before use) and then three times with an ethanol-toluene mixture (1 : 4). Wash the precipitate and paper back into the original beaker with water, and dilute to about 100 ml. Heat until the precipitate dissolves. Add 5 ml. 0·5 per cent solution of eosin Y and titrate with 0·03 N silver nitrate solution. Compare the titration value obtained with that obtained in a standard test using pure cadmium.

[1] T. L. Thompson, *Ind. Eng. Chem. Anal.*, 1941, **13**, 164.

3. TITRIMETRIC DETERMINATION OF CALCIUM BY PERMANGANATE

(a) METHOD OF McCOMAS AND RIEMANN

According to McComas and Riemann,[1] 20 to 160 mg. calcium can be determined in limestone without interference from magnesium and other elements, by precipitating slowly as oxalate, digesting the precipitate for 5 minutes in a hot, strongly acid solution, cooling rapidly, adjusting the pH to 3·7 by a formate buffer, and finally digesting for 30 minutes at 25°C. This procedure prevents post-precipitation of magnesium and at the same time gives a readily filterable precipitate. The precipitate may then be redissolved and titrated with standard permanganate.[2] Alternatively it may be determined gravimetrically by drying at 110°C. for 2 hours and weighing as monohydrate, by igniting for one and a half hours at 485°C. and weighing as carbonate, or by igniting strongly in platinum to constant weight, using a Meker burner, and weighing as oxide.

Under the conditions of precipitation described, aluminium, iron, manganese, sodium and titanium in amounts greater than normally found in limestone do not coprecipitate. Post-precipitation of magnesium, normally a serious drawback in the determination of calcium in a limestone, does not occur to a serious extent. Silica should first be removed if a gravimetric finish is chosen. Phosphate will produce an error if the finish is titrimetric, because of some coprecipitation of calcium phosphate. This error, however, is very small in a gravimetric finish.

Procedure

Titrimetric determination of calcium in limestone. Ignite about 300 mg. limestone, weighed accurately, in a platinum crucible, adding 25 mg. sodium carbonate in the case of an argillaceous limestone. Cool, and add cautiously 2 ml. water and 2 ml. 6 N hydrochloric acid. Stir, and transfer to a 400-ml. beaker. Add a further 2 ml. water and 2 ml. acid to the crucible and transfer this to the beaker. Finally wash the crucible thoroughly and add the washings to the beaker. Warm to dissolve all but the silica. Dilute to 100 ml. and add a slight excess of bromine. Add 10 drops 0·1 per cent bromophenol blue, and follow this with 6 M ammonia, drop by drop, until the solution turns yellow or ferric hydroxide precipitates. Add 12 ml. 2 M formic acid, and dilute to 140 ml. Heat the solution to 95°C. and add 20 ml. 0·5 M oxalic acid over 85 seconds, stirring continuously. Digest the precipitate for 5 minutes at 85°–90°C. and then transfer the beaker to a large beaker of cold water. At the end of 15 minutes the solution should have cooled to 25°C. Add 34 ml. 2 M ammonium formate solution over 2½ minutes, stirring continuously. Digest for 30 minutes at 25°C.

Prepare an asbestos mat supported in an ordinary filter funnel on a Witt plate. Filter the bulk of the supernatant liquid through this, wash the precipitate three times by decantation with 20-ml. portions of a wash solution containing 0·02 mole ammonium oxalate and 0·01 mole oxalic acid per litre. Transfer the precipitate to the filter with a further 150 ml. wash solution, and wash with three 8-ml. portions of water.

Transfer the precipitate, disc and asbestos to a 400-ml. beaker. Dilute 12·5 ml. concentrated sulphuric acid to 250 ml., boil for 10–15 minutes, and cool to 25°–30° C. Add this to the beaker, and stir till the precipitate is completely dissolved (5–10 minutes). Add the bulk of the 0·1 N permanganate required for neutralisation at the rate of 25 ml. per minute, stirring slowly. Allow the solution to stand for 45 seconds so that the pink colour disappears. Warm to 55°–60° C. and complete the titration by adding standard permanganate solution dropwise, allowing each drop to become decolorised, until a faint pink colour persists for 30 seconds. Determine an indicator blank on another portion of the diluted sulphuric acid treated in a similar fashion.

1 ml. 0·1 N potassium permanganate≡2·804 mg. CaO.

≡2·004 mg. Ca.

[1] W. H. McComas and W. Riemann, *Ind. Eng. Chem. Anal.*, 1942, **14**, 929.
[2] R. M. Fowler and H. A. Bright, *J. Res. Nat. Bur. Stds.*, 1935, **15**, 493.

(b) Method of Lingane

Lingane[1] has utilised a single precipitation of calcium oxalate from acid solution, followed by titrimetric determination with permanganate, for the direct determination of calcium in the presence of appreciable amounts of aluminium, iron, magnesium, phosphate and silica. He finds that for this purpose the precipitation procedure originally devised by Richards and his co-workers for the gravimetric determination,[2] which consists in slowly neutralising a strongly acid solution with ammonia to a final pH of 3·5–4·5 is more reliable than that of McComas and Riemann (p. 205).

In 0·4 g. samples of dolomite containing 30–40 per cent calcium oxide the separation from magnesium is satisfactory. Amounts of aluminium and iron up to 100 mg. oxide, and of phosphate up to 100 mg. phosphorus pentoxide have no appreciable effect. Titanium oxide and manganous oxide may be present only up to amounts of about 2 mg., but calcareous materials do not usually have a content of these oxides in excess of this amount. The method may consequently be applied successfully to limestones, Portland cement, phosphate rocks and similar materials. The permissible amounts of foreign elements are considerably in excess of the amounts allowed by the method of McComas and Riemann.

The accuracy is high, and duplicate determinations agreeing within 1 part in 1000 may be completed within 2 hours.

Procedure

Titrimetric determination of calcium in limestone. Weigh a 0·4-g. sample into a 250-ml. wide-mouthed conical flask. Add 5 ml. water and 10 ml. 12 N hydrochloric acid. Warm for a few minutes on the steam bath to complete decomposition. Dilute to 50 ml. with water and heat almost to boiling. Add through a filter paper 100 ml. aqueous 5 per cent ammonium oxalate monohydrate solution, previously heated to 90°C. Follow this with a few drops 0·1 per cent methyl orange indicator solution. Adjust the temperature, if necessary, to 80°C., and add 7 N ammonia dropwise over 5–10 minutes, with constant and thorough shaking, until the methyl orange turns a pinkish-yellow identical with that obtained by adding the same amount of indicator to an equal volume of 0·1 M potassium hydrogen phthalate. Set aside for 20–30 minutes. A longer time than this may produce post-precipitation of magnesium.

Filter on to an asbestos mat in a Gooch crucible, and wash thoroughly with a maximum of 100 ml. ice-cold water applied in 8–10 small portions. Wash down the outside of the Gooch crucible and place it in the original conical flask. Add 100 ml. water and 5–6 ml. 36 N sulphuric acid, and warm to about 90°C.

Add from a burette 2–3 drops 0·1 N potassium permanganate solution, wait till the solution is decolorised, and then proceed with the titration in the usual manner, ensuring that the temperature does not fall below 60°C.

Standardise the potassium permanganate solution frequently against sodium oxalate dried at 110°C. Use an amount of oxalate sufficient to give a titration figure within 2 ml. of the titration figure for the determinations, thus avoiding a blank correction.

[1] J. J. Lingane, *Ind. Eng. Chem. Anal.*, 1945, **17**, 39.
[2] T. W. Richards, C. F. McCaffrey and H. Bisbee, *Proc. Amer. Acad. Sci.*, 1901, **36**, 377.

(c) METHOD OF PROČKE AND MICHAL

Pročke and Michal[1] have studied the method of Fox[2] for the determination of calcium and magnesium in an attempt to improve the accuracy, and eliminate the need for empirical corrections. They find that low results for magnesium are caused by using too great an excess of oxalic acid to precipitate calcium, complex magnesium oxalates being formed. The amount of ammonium chloride added must also be limited, otherwise low results are obtained both for calcium and magnesium. When ammonium arsenate is added after all the other reagents, low results are obtained for magnesium, probably because of precipitation of $Mg_3(AsO_4)_2$ together with $MgNH_4AsO_4$.

Precipitation in the presence of tartaric acid fails to prevent completely the interference of aluminium, high results being obtained. Although the effect of iron was not investigated, it would probably interfere even more than aluminium because of the oxidation of iodide by iron-III. Aluminium and iron must, therefore, be separated if present.

When precautions are taken to avoid the above sources of error, satisfactory results are obtained without recourse to the use of empirical factors.

The method has been applied to the analysis of dolomite, calcite and magnesite.

Procedure

Titrimetric determination of calcium in dolomite. Dissolve 0·15–0·2 g. in 3 ml. dilute hydrochloric acid in a flask, add 10 ml. 10 per cent ammonium chloride solution, 50 ml. 2 per cent ammonium arsenate solution (prepared by neutralising pure arsenic acid with ammonia and diluting to give a 2 per cent solution of $(NH_4)_2HAsO_4$) and 0·2–0·3 g. oxalic acid. Dilute to about 150 ml., add methyl orange indicator, and heat to boiling. Add ammonia solution (1 : 9) slowly with constant stirring until the indicator turns yellow. After a short interval add more ammonia solution until the precipitate agglomerates and the solution becomes clear. When the precipitate is granular, calcium preponderates, whereas when it is curdy magnesium is in excess. Digest for at least 1 hour on a water-bath, add 20 ml. ammonia solution (1 : 4), allow to cool, shake slightly and set aside overnight.

15

Filter, using a filter stick, and wash five times with 10-ml. portions of 1 : 40 ammonia solution. Dissolve the precipitate in 50 ml. 4 N sulphuric acid, taking care to include precipitate adhering to the filter stick. Heat to 70° C. and titrate with 0·1 N potassium permanganate.

Titrimetric determination of magnesium in dolomite. Cool the solution from the calcium determination, add 2 g. solid potassium iodide and follow this by sulphuric acid (1 : 1) until a precipitate of arsenious iodide appears. Cover the flask with a watch-glass and set aside in the dark for 10 minutes. Titrate very slowly with 0·1 N sodium thiosulphate solution, stirring constantly until the solution is decolorised.

1 ml. 0·1 N sodium thiosulphate≡2·016 mg. MgO.

[1] O. Pročke and J. Michal, *Coll. Czech. Chem. Comm.*, 1938, **10**, 20.
[2] P. J. Fox, *Ind. Eng. Chem.*, 1913, **5**, 910.

(*d*) BY UREA HYDROLYSIS

The urea hydrolysis method of Willard and Furman[1] for precipitating calcium oxalate has been subjected to further study by Ingols and Murray.[2] The method is extremely attractive. Urea is added to the slightly acid solution containing calcium and oxalate ions, and when the solution is boiled the urea decomposes slowly according to the equation:

$$(NH_2)_2CO + H_2O \longrightarrow 2NH_3 + CO_2.$$

The slow raising of the pH results in the formation of very large crystals, as can be shown by comparing photomicrographs of calcium oxalate precipitated in this manner with those of crystals precipitated by the conventional method. The precipitate may be filtered shortly after precipitation, and contamination is slight.

When sulphate is present both this and the normal method of precipitation yield high results, but the error is less by the urea hydrolysis method. The error is not reduced by continued washing or by dilution of the solution prior to precipitation. A second precipitation by the urea hydrolysis method gives exact results, however, whereas the conventional method still shows appreciable error.

The newer method gives better results if magnesium is present since 150 p.p.m. do not co-precipitate unless sulphate is also present. If as little as 7 p.p.m. of aluminium is present it is co-precipitated completely by the conventional method, whereas, under the same conditions, the urea hydrolysis method yields better, although not precise results. Theoretical results can be obtained by adding tartrate (ten times the equivalent of the aluminium) and allowing 30 minutes for completion of the reaction before proceeding with the precipitation of calcium oxalate.

Procedure

Titrimetric determination of calcium. Add a few drops methyl red to 100 ml. calcium-containing solution, followed by 2·4 ml. hydrochloric acid (1 : 1). This should produce a pH of 1·0. Add 15 ml. saturated ammonium oxalate solution and 10 g. of urea, and stir thoroughly. There should be no precipitation at this stage.

The urea must be added as solid or as a freshly prepared solution, since a 24-hour old solution contains sufficient ammonia to cause immediate precipitation. Heat on a hot plate until the methyl red changes colour at pH 5·0 (generally 15 minutes). The precipitate is then ready for filtration.

A coarse qualitative filter paper may be used for the filtration, or filtration may be carried out through a small filter paper supported in a Gooch crucible. The precipitate is then titrated in the usual way with standard potassium permanganate solution.

[1] H. H. Willard and N. H. Furman, *Elementary Quantitative Analysis*, 3rd Ed., New York, 1940, p. 397.
[2] R. S. Ingols and P. E. Murray, *Analyt. Chem.*, 1949, **21**, 525.

4. TITRIMETRIC DETERMINATION OF CALCIUM BY HEXANITRATOCERATE

Ceric sulphate as an oxidising agent for the direct titration of oxalate was proposed in 1934.[1] Ammonium hexanitratocerate, $Ce(NH_4)_2(NO_3)_6$, was recommended as an oxidimetric reagent in 1936,[2] and has found considerable application. Kochiakin and Fox[3] claim to have found that very stable 0·01 N solutions are obtained by making up the reagent in perchloric acid solution, and that Setopaline C is a very satisfactory indicator for titration against oxalate solutions.

The method has been adapted to the determination of calcium precipitated as calcium oxalate.

Solutions required

Standard ammonium hexanitratocerate solution. Dissolve a weighed amount of the AnalaR salt directly in N perchloric acid solution, and make up to the required amount, maintaining the concentration of perchloric acid at N. Do not use heat to hasten solution, as this causes decomposition and results in the formation of a large amount of a fine white precipitate which interferes with the titration. Store the solution in a black glass bottle or in the dark, and standardise it against sodium oxalate. A solution prepared and stored in this way decreases in strength less than 2·5 per cent in 90 days.

Setopaline C indicator solution. Suspend 50 mg. Setopaline C in 100 ml. distilled water, and warm to cause solution. The warm solution must be used for the titration, since solid is deposited on cooling and this solid is immediately oxidised by the reagent.

Procedure

Titrimetric determination of calcium oxalate. Wash the precipitated calcium oxalate thoroughly and dissolve in N perchloric acid solution. Add 1 ml. warm indicator solution for 25 ml. solution, and titrate with the standard cerate solution until a sharp colour change from yellow to salmon-pink occurs. The salmon-pink changes in about 15 seconds to bronze.

1 ml. 0·1 N ammonium hexanitratocerate≡2·004 mg. Ca.

[1] F. Rappaport and D. Rappaport, *Mikrochem. Mikrochim. Acta*, 1934, **15**, 107.
[2] G. F. Smith, V. R. Sullivan and G. Frank, *Ind. Eng. Chem. Anal.*, 1936, **8**, 449.
[3] C. D. Kochiakin and R. P. Fox, *ibid.*, 1944, **16**, 762.

5. TITRIMETRIC DETERMINATION OF CALCIUM BY SODIUM FLUORIDE

When calcium is titrated with sodium fluoride in the presence of ferric thiocyanate, the red colour of the latter disappears at the equivalence point. The solubility product of calcium fluoride is 4×10^{-11}, hence the concentration of the calcium at the equivalence point is 2×10^{-4} and that of the fluoride is 4×10^{-4}, a value which is too high to permit a sharp jump in pH at the equivalence point. By adding ethanol the solubility can be decreased, and the colour change becomes sharper.[1] One or two drops of $0 \cdot 1$ M ferric chloride are sufficient, but a large excess of ammonium thiocyanate is necessary.

The optimum pH range is pH $2 \cdot 5$–$3 \cdot 5$. At higher values the colour intensity of ferric thiocyanate decreases, and at lower values calcium fluoride begins to dissolve.

Procedure

Titrimetric determination of calcium. Adjust the solution to have a volume of 10–20 ml. and to contain 50–100 mg. calcium. Add $0 \cdot 1$ ml. $0 \cdot 1$ M ferric chloride solution, 4 ml. 60 per cent ammonium thiocyanate solution, and enough ethanol to double the volume. Titrate with $0 \cdot 5$ M sodium fluoride, using a 10-ml. burette graduated in 1/100 ml., until the red colour disappears. Apply a correction for consumption of fluoride by the indicator.

$$1 \text{ ml. } 0 \cdot 1 \text{ M sodium fluoride} \equiv 2 \cdot 004 \text{ mg. Ca.}$$

Titrimetric determination of aluminium. Carry out the determination precisely as described for calcium.

$$1 \text{ ml. } 0 \cdot 1 \text{ M sodium fluoride} \equiv 0 \cdot 899 \text{ mg. Al.}$$

Standardisation of sodium fluoride solution. Standardise against pure potassium alum. Conduct the titration exactly as described, but saturate the solution with sodium chloride prior to the determination. Since potassium alum is but slightly soluble in ethanol, high potassium concentrations should be avoided. For the same reason, sodium fluoride and ammonium thiocyanate should not be replaced by the corresponding potassium salts.

[1] A. Ringbom and B. Merikanto, *Acta Chem. Scand.*, 1949, **3**, 29.

6. GRAVIMETRIC DETERMINATION OF CALCIUM

(*a*) As Calcium Oxalate Monohydrate

Calcium is usually weighed as the oxide, the carbonate or, less occasionally, the sulphate. The oxide is hygroscopic; the carbonate, formed by decomposition of the oxalate, requires very strict temperature control; the sulphate requires very careful handling.

Methods using calcium oxalate monohydrate as the weighing form (p. 206) which appear at first sight to offer advantages, have been recommended. However, the use of calcium oxalate monohydrate can introduce serious errors, and the method has accordingly received little attention.

A careful study of the sources of error has been made by Kolthoff and Sandell,[1] who found that when calcium oxalate is precipitated in the usual way from neutral, acid or ammoniacal medium it retains foreign water after air-drying at room temperature. This foreign water is not removed even after drying over concentrated sulphuric acid. Even when dried at 105° C. or above the foreign water is still retained, although in small amounts, and in these conditions there is the risk of loss of hydrate water.

If the precipitation is carried out in the cold in presence of acetic acid (to prevent basic oxalate formation) by the sudden addition of ammonium oxalate followed by 20 hours' digestion near the boiling point, the product is only slightly hygroscopic, and there is little occluded water retained after air-drying. High proportions of the di- and trihydrates are formed, but these are unstable in hot solution. Relatively perfect crystals of the monohydrate are formed, retaining only slight amounts of foreign water. The crystals are small, however, and slow down the filtration.

Precipitation by Willard and Furman's urea method (p. 210) gives better results. The amount of foreign water is small. Precipitation must not be too rapid. It was found that 10–15 g. urea in 200 ml. solution containing 5 ml. concentrated hydrochloric acid, when hydrolysed over 1–2 hours, gave results which were 0·2–0·5 per cent high after air-drying. When 5 g. urea was used and the hydrolysis allowed to proceed overnight at 80°–90° C., the error was +0·2 per cent or less for 0·25–0·5 g. calcium as the carbonate. For 0·05–0·1 g. the positive errors increased owing to coprecipitation of the oxalate. When dried at 105° C. the results are closer to theory than using air-drying, but there is possible loss of hydrate water. The coprecipitation of sodium is marked, and unless only small amounts are present the calcium oxalate should be reprecipitated. Magnesium in small amounts does not interfere.

Procedure

Gravimetric determination of calcium. Dilute to 175–200 ml. the solution containing 0·2–0·5 g. calcium expressed as the carbonate.

Add 5 ml. concentrated hydrochloric acid and heat nearly to boiling. Add 1·0 g. ammonium oxalate monohydrate dissolved in 20 ml. hot water. No precipitate should form. Add 5·0 g. urea dissolved in a small amount of water, and leave at 80°–90° C. overnight, when the solution should be alkaline to methyl orange. Cool, filter on a Gooch porous-porcelain or sintered-glass crucible which should have been weighed after standing 10–15 minutes in the air. Wash the precipitate with a small amount of cold water, then with three or four 2-ml. portions of acetone. Draw air through the filter for 5–10 minutes, and weigh. If the sodium or magnesium content is appreciable, precipitate first by the conventional method. After washing with ammonium oxalate solution, redissolve, and reprecipitate as described above. Only 0·2–0·3 g. ammonium oxalate is then required.

$$\text{mg. ppt.} \times 0\cdot2743 = \text{mg. Ca.}$$

[1] I. M. Kolthoff and E. B. Sandell, *Ind. Eng. Chem. Anal.*, 1939, **11**, 90.

(b) As Sulphate

In order to overcome the incomplete precipitation of calcium as oxalate in the presence of high concentrations of magnesium, Caley and Elving[1] have proposed precipitation as sulphate in a solution of high methanol content. The precipitation is complete enough to give quantitative results if the concentration of methanol is at least 90 per cent, and the solubility of magnesium sulphate is not reduced to the same extent as by the presence of high concentrations of ethanol. Consequently calcium can be separated from a large amount of magnesium.

The calcium sulphate precipitates as a variable mixture of hemi-hydrate and dihydrate. This is ignited and weighed as the anhydrous salt.

The precise method of precipitation has a critical effect on the physical nature of the precipitate, and hence on the filtering time and on the washing efficiency, although it appears to have no effect on completeness of precipitation. Optimum results are obtained by adding sulphuric acid to the aqueous solution, and then precipitating the calcium sulphate by slow addition of methanol up to the required concentration. Two procedures have been devised, and a selection of the appropriate method is made according to Table IX.

TABLE IX

Calcium present	Foreign ions	Method	Total volume of solution
Up to 5 mg.		I	50 ml.
5–100 mg.		I	100 ml.
About 100 mg.	Low concentration	I	100–200 ml.
About 100 mg.	High concentration	II	200 ml.
More than 100 mg.		II	200 ml.

As the calcium sulphate precipitate produced by the methanol is very voluminous a large volume of solution is required where the calcium content is high. To ensure the proper methanol content in the final solution at least 99 per cent methanol should be used, although it is not essential to use the absolute grade. As long as at least twice the amount of sulphuric acid required for complete precipitation of the calcium is present, the amount of this reagent is not critical, and a considerable excess may be used. However, in the presence of large amounts of foreign ions too great an excess of sulphuric acid should be avoided because of the possibility of co-precipitation.

When calcium alone is present the precipitation is complete in a short time (15–30 minutes). If there is less than 1 mg. calcium somewhat longer waiting time may be necessary. This is also the case where magnesium is present. If the amount of calcium is large the retardation of precipitation by magnesium is not marked, but where the amount of calcium is low (1 mg. calcium to 100 mg. magnesium) 4 hours are necessary for complete precipitation.

When the amount of calcium alone is about 200 mg. low results are obtained, so that the amount of sample should be restricted to keep the calcium content below this amount.

In addition to magnesium in considerable amounts (Mg : Ca 200: 1) small amounts of aluminium, iron and manganese (3 per cent) and very small amounts of potassium and sodium (0·5 per cent) may also be present.

Ammonium salts, barium, lead and strontium should be absent. Acids other than sulphuric acid should be absent, or should only be present in low concentration, since they increase the solubility of the calcium sulphate. Hence the sample is first transformed to a neutral solution of the nitrate, chloride or perchlorate.

Procedures

Gravimetric determination of calcium. Method I. Evaporate the solution to dryness and redissolve in 4·5, 9 or 18 ml. water for a final volume of 50, 100 or 200 ml. solution respectively. Add 0·5, 1·0 or 2·0 ml. 9 N sulphuric acid, and slowly add, with constant stirring, the correct amount (45, 90 or 180 ml.) of methanol to bring the aqueous solution to the desired final volume. Set aside for the time indicated by the nature of the sample as described above. Filter through a porous-porcelain filter crucible. Wash with 90 per cent methanol, first by decantation, then by stirring up the precipitate with the wash solution and allowing a few minutes' contact before filtration. Depending on the amount of calcium and other ions present, 30–100 ml. wash liquid will be required.

Dry the precipitate at 100°–110° C. for 30–45 minutes. Ignite in an electrically heated muffle furnace for 30–45 minutes at 400°–450° C., cool and weigh.

$$\text{mg. ppt.} \times 0.2594 \equiv \text{mg. Ca.}$$

Gravimetric determination of calcium. Method II. Add 1·0 or 2·0 ml. 9 N sulphuric acid to the solution, and evaporate to a volume of 5 ml. Add 15 ml. water, and precipitate the calcium sulphate by slow addition, with constant stirring, of 180 ml. methanol. Complete the determination as in Method I.

Determination of calcium in burned magnesite. Transfer 1 g. sample to a 250-ml. beaker, add 5 ml. water and 10 ml. 60 per cent perchloric acid. Heat on the steam bath till solution is complete or only a residue of silica remains. Fume off the perchloric acid, cool, and redissolve the residue in 20 ml. water. Filter off the silica, and wash with 50 ml. hot water in small portions, adding the washings to the filtrate.

Evaporate the filtrate and washings to 9 ml., add 1 ml. 9 N sulphuric acid, and precipitate by adding, with constant stirring, 90 ml. methanol. Set aside for at least 1 hour (several hours if the amount of precipitate is small), filter as in the general procedure, wash with a total volume of 50 ml. 90 per cent methanol, and complete the determination as already described.

[1] E. R. Caley and P. J. Elving, *Ind. Eng. Chem. Anal.*, 1938, **10**, 264.

(c) As Tungstate

Peltier and Duval[1] have examined several methods for the determination of calcium in order to find a method suitable in the presence of at least 20 times the weight of magnesium. The methods examined were:

1. Precipitation as calcium tungstate.
2. Precipitation as oxalate in the presence of

   (a) glycerol.
   (b) formic acid.
   (c) acetic acid.

3. Precipitation as oxalate by the classical methods.
4. Precipitation as oxalate, but using instead of ammonia,

   (a) pyridine.
   (b) aniline.
   (c) urea.
   (d) antipyrine.
   (e) hexamethylenetetramine.

5. Precipitation as sulphate in the presence of ethanol and reprecipitation as oxalate.
6. Precipitation as molybdate.
7. Precipitation of the magnesium with 8-hydroxyquinoline followed by the precipitation of the calcium in the filtrate as oxalate.
8. Precipitation as the complex triple nitrite of potassium, calcium and nickel, $K_2Ca[Ni(NO_2)_6] . 3H_2O$.
9. Precipitation as the tartrate, iodate, sulphite or picrolonate.

They claim that the only satisfactory methods are 1, 2(a), 2(b) and 5, and that of these, Method I, the tungstate method, is best, since it is rapid and accurate. The other methods are complicated and time-consuming.

Procedure

Gravimetric determination of calcium. Dilute the solution containing about 0·035 g. calcium, to about 70 ml. and adjust to pH 7·0–8·0. Heat to 80°C., add 2 ml. 19 per cent sodium tungstate solution and stir without touching the sides of the vessel until precipitation is complete. Allow to cool, and after 20 minutes filter on an X3 sintered-glass filter crucible. Wash with 20 ml. hot water. Dry for 1 hour at 110°C., cool, and weigh.

In addition to calcium and magnesium only the alkali metals should be present. Large amounts of ammonium chloride prevent the precipitation.

$$\text{mg. ppt.} \times 0.1947 \equiv \text{mg. CaO.}$$
$$\text{mg. ppt.} \times 0.1391 \equiv \text{mg. Ca.}$$

[1] S. Peltier and C. Duval, *Anal. Chim. Acta*, 1947, **1**, 408.

7. TITRIMETRIC DETERMINATION OF CHLORATE

Conventional methods for the determination of chlorate consist either of addition of an excess of a reducing agent and titration of the excess, or the determination of iodine liberated from a ferrous sulphate-potassium iodide system. A highly accurate and precise method has been devised[1] in which the chlorate is reduced by excess of ferrous ion, using ammonium molybdate as catalyst, and the excess ferrous ion is titrated with potassium dichromate. The method may be applied to the determination of chlorate in the cell liquor produced in chlorine manufacture.

Solutions required

Buffer solution. Add 250 ml. syrupy phosphoric acid to 1 litre 4 M sodium acetate solution.

Indicator solution. Dissolve 0.30 g. barium diphenylamine sulphonate in 100 ml. water. Add 0.5 g. sodium sulphate and filter free from precipitated barium sulphate.

Procedure

Titrimetric determination of chlorate. Measure out 10 ml. of the solution and if necessary neutralise this to phenolphthalein with N hydrochloric acid. Add 10 ml. 0.25 N ferrous ammonium sulphate solution, 3 drops 10 per cent ammonium molybdate solution and 40 ml. concentrated hydrochloric acid. The reaction is complete after 1 minute.

Add 20 ml. buffer solution, dilute to 200 ml. with water, and add 2 drops indicator solution. Titrate with 0.1 N dichromate solution to the purple end-point. Deduct 0.025 ml. as indicator correction.

1 ml. 0.1 N potassium dichromate $\equiv 1.391$ mg. ClO_3^-.

[1] A. J. Boyle, V. V. Hughey and C. C. Casto, *Ind. Eng. Chem. Anal.*, 1944, **16**, 370.

8. TITRIMETRIC DETERMINATION OF CHLORIDE

Diphenylcarbazide disodium disulphonate, or the disodium salt of
1 : 5-bis-(1-phenyl-4-sulphonic acid)-carbohydrazide,

is not subject to the gradual colour development shown by either
diphenylcarbazide or diphenylcarbazone in the mercurimetric
titration of chloride, but gives a very sharp colour change at the end-
point. This change is even better if, just before the titration, an
oxidising agent (which is conveniently iron-III) is added, to convert
the compound to diphenylcarbazone disulphonate; the colour change,
which is from yellow-orange to violet, can also be accentuated by
screening with nickel solution.[1]

Solutions required

Standard mercuric nitrate solution. Dissolve 18 g. mercuric
nitrate in 100 ml. 0·1 N nitric acid. Filter if necessary, and dilute to
1 litre. Adjust the pH if necessary to $2 \pm 0·1$, and standardise against
pure sodium chloride.

Indicator solution. 2 per cent aqueous.

Nitric acid. 0·4 N prepared from colourless nitric acid (sp. gr.
1·42).

Oxidising-screening solution. Dissolve 20·5 g. ferric nitrate,
$Fe(NO_3)_3 . 9H_2O$, in water, add 25 ml. nitric acid (sp. gr. 1·42) and
dilute to 1 litre. Dissolve 58 g. nickel nitrate, $Ni(NO_3)_2 . 6H_2O$, in
warm water and dilute to 50 ml. Mix this with 25 ml. of the ferric
nitrate solution, and dilute to 100 ml.

Procedure

Titrimetric determination of chloride. Adjust the volume of chloride
solution, containing sufficient chloride to give a titration figure of at
least 15 ml., to 50–60 ml., so that the volume at the end of the titration
will be 75–100 ml. Adjust to pH 2·0 by adding 2 ml. 0·4 N nitric acid,
and using either indicator paper or a glass electrode if salts likely to
buffer the solution are present. Transfer 10 ml. water, 5 drops
(0·25 ml.) indicator solution and 1 ml. iron-nickel solution to a 25-ml.

cylinder. Mix well to ensure uniform formation of the diphenylcarbazone disulphonate, and then add this to the test solution. Titrate
with standard mercuric nitrate solution to a neutral grey which
immediately precedes the end-point. Continue the titration with split
drops to the appearance of a purple colour. One drop of mercuric
nitrate solution turns the solution from green to grey, and a fraction
of a drop is sufficient to complete the titration to the purple colour.

$$1 \text{ ml. } 0.1 \text{ N mercuric nitrate } 3.546 \equiv \text{mg. Cl}^-.$$

Synthesis of diphenylcarbazide disodium disulphonate[2]

Dissolve 0.3 mole (63 g.) sodium salt of phenylhydrazine p-sul-
phonic acid and 15 g. sodium carbonate in 400 ml. distilled water.
Heat to 50° C. and saturate with phosgene, maintaining efficient
saturation for 3–4 hours. Adjust the solution, by now probably
slightly acid, to pH 1.0 with hydrochloric acid. Filter, and discard
any precipitate.

Pre-wash an alumina column (3×75 cm. 80–100 mesh) with water
and then with 0.5 N hydrochloric acid, and pass the solution through
the column at 5 ml. per minute. The product is absorbed as a colourless band. Wash the column with water until the filtrate is as free
from chloride as possible. Wash the product through the column with
0.25 M sodium hydroxide solution. Discard the first 200–300 ml.
eluate, which is almost colourless and contains very little of the
product. Collect 400 ml. yellowish-orange eluate (pH 4.0–7.0) which
contains the bulk of the product. Discard the final red solution
containing a small amount of the alkaline form together with excess
sodium hydroxide.

Concentrate the yellowish-orange solution under reduced pressure,
in an atmosphere of nitrogen, at 40°–50° C., thus reducing oxidation
of the carbazone to a minimum. Precipitate the product with 95 per
cent ethanol, filter, redissolve in a small amount of water, and reprecipitate with ethanol. Filter, dry *in vacuo* at room temperature,
and store in a desiccator over a saturated solution of calcium chloride
to maintain 30 per cent relative humidity. The product contains 3
molecules of water of crystallisation, which are lost on drying at
105°–110° C.

Yield of purified material about 10–20 per cent.

[1] J. S. Parsons and J. H. Yoe, *Anal. Chim. Acta*, 1952, 6, 217.
[2] *Idem, J. Amer. Chem. Soc.*, 1951, 73, 5482.

9. TITRIMETRIC DETERMINATION OF COPPER

(a) IODOMETRICALLY

Copper is frequently determined by using standard thiosulphate solution to titrate the tri-iodide formed from added potassium iodide. The reactions involved are:

$$2Cu^{++} + 5I^- \longrightarrow Cu_2I_2 + I_3^-. \tag{1}$$

$$I_3^- + 2S_2O_3^= \longrightarrow S_4O_6^= + 3I^-. \tag{2}$$

A moderate concentration of hydrogen ions increases the rate of the reaction and prevents undesirable side reactions. As a rule the titration is carried out in buffered acetic acid solution, although Whitehead and Miller[1] have claimed that a considerable variation in the concentration of hydrogen ions has little effect on the results.

The method is rapid, but suffers from a number of disadvantages. The excess iodide required is rather expensive. The relative proportions of copper-II and added iodide are quite critical, optimum results being obtained when 3 g. potassium iodide and a final volume of 100 ml. are used in the estimation of 0·2 g. copper.[2] More serious is the fact that reaction (1) is reversible to some extent, owing to (a) the rather high solubility of cuprous iodide, and (b) adsorption of iodine by cuprous iodide, which is apparent by the fact that the precipitate is distinctly coloured at the end-point.

As a consequence of this non-stoicheiometric relation between copper and thiosulphate, the method gives low results unless the thiosulphate is standardised under conditions closely approximating to those of the actual determinations.

Bruhns[3] showed that it is possible to reduce the amount of iodide by first adding potassium or ammonium thiocyanate. The solution darkens somewhat on this addition, but no precipitate forms since cupric thiocyanate is soluble. On addition of potassium iodide cuprous iodide is precipitated as in the original determination. But since cuprous thiocyanate is much less soluble (0·5 mg. per l.) than cuprous iodide (8·0 mg. per l.) there is an immediate conversion:

$$Cu_2I_2 + 2CNS^- \longrightarrow Cu_2(CNS)_2 + 2I^-.$$

The greater insolubility of the cuprous thiocyanate results in a more complete reduction of copper-II ions. In addition, iodine is not adsorbed on the thiocyanate precipitate, which is practically white, and therefore a much sharper end-point is obtained. The amount of iodide added can be reduced to about 2 g. per 0·16 g. copper, a suitable amount of potassium thiocyanate for such a titration being 1·25–1·5 g.

Other procedures are also possible. The copper solution, with added potassium iodide, may be titrated with a mixed thiosulphate-thiocyanate solution. The copper solution may be titrated directly with a mixed potassium iodide-thiosulphate-thiocyanate solution. Or a mixture of solid thiocyanate and iodide may be added, followed by immediate titration with thiosulphate. The last of these procedures appears to give the most satisfactory results.

Critical examination of the method[3] shows that unreliable results are still obtained, because of a side reaction in which thiocyanogen is liberated, which subsequently reacts with the free iodine.

$$2Cu(CNS)_2 \longrightarrow Cu_2(CNS)_2 + (CNS)_2.$$

Further work on the method[4] showed that this undesirable reaction can be practically eliminated by addition of the thiocyanate at a late stage in the titration. By this method the surface layers at least of the precipitate are converted from cuprous iodide to cuprous thiocyanate, and adsorbed iodine is liberated, so that the precipitate immediately becomes white. Liberation of the iodine is indicated by a marked deepening of the colour of the starch-iodide complex. About 2–3 g. potassium iodide and 2 g. ammonium thiocyanate are recommended.

Using this method, results which are comparable in accuracy with the best iodometric methods are obtained. The authors claim that a wide variation in dilution has little effect, and that in accordance with the findings of Whitehead and Miller on the original method, buffered and unbuffered solutions show no measurable difference. Although it is claimed that complete removal of nitric acid is unnecessary, elimination of nitrous fumes by boiling with urea is usually carried out.

One of the claims for the thiocyanate method has always been that it economises in potassium iodide, but the tendency has been to overlook this in the experimental instructions for the method. The amount of potassium iodide recommended in standard textbooks (e.g., 3 g. per 0·25 g. copper[5] : 2 g. per 25 ml. 0·1 N copper solution[6]) is actually closely comparable with the amount found by Herschkowitz[2] to give optimum results in the simple iodometric estimation of copper by thiosulphate without added thiocyanate.

A more recent investigation has confirmed[7] that it is possible to economise considerably in the amount of potassium iodide used without decreasing the accuracy of the improved procedure. Apart from a reduction in the amount of potassium iodide, no significant alteration in procedure is necessary.

Procedure

Titrimetric determination of copper. If the copper has been dissolved in nitric acid remove the nitric acid by evaporation with sulphuric acid, and then redissolve in water to give an approximate concentration of 0·05–0·1 M. The hydrogen-ion concentration need not be controlled by buffering, so long as the resulting solution is only moderately acid.

To 25 ml. of the solution add 0·5–0·6 g. potassium iodide, either as solid or as 10 per cent solution. Titrate with standard sodium thiosulphate solution until the iodine colour fades. Add starch solution and continue titration till the blue colour just fades. Add 10 ml. 10 per cent solution of ammonium or potassium thiocyanate, which will revive the blue colour. Complete the titration as rapidly as possible. The end-point should now be quite sharp, and there should be no reappearance of the blue starch-iodine colour on standing.

1 ml. 0·1 N sodium thiosulphate≡6·357 mg. Cu.

[1] T. H. Whitehead and H. S. Miller, *Ind. Eng. Chem. Anal.*, 1933, **5**, 15.
[2] M. Herschkowitz, *Z. anorg. Chem.*, 1925, **146**, 132.
[3] G. Bruhns, *Chem.-Ztg.*, 1918, **42**, 301: *Z. anal. Chem.*, 1919, **58**, 128: 1920, **59**, 337.
[4] *Idem, Centralblatt Zuckerind.*, 1918, **26**, 354: D. Krüger and E. Tschirch, *Z. anal. Chem.*, 1934, **97**, 161.
[5] A. I. Vogel, *Quantitative Inorganic Analysis*, London, 1951, pp. 337–338.
[6] G. Fowles, *Volumetric Analysis*, London, 1932, p. 145.
[7] C. C. Oglethorpe and C. G. Smith, *Analyst*, 1943, **68**, 325.

(b) By Dichromate

Copper-II is frequently determined by reduction and precipitation as cuprous thiocyanate, the determination being completed in a variety of ways. The most usual finishes are:

1. Direct gravimetric determination of the precipitated cuprous thiocyanate.

2. Determination by silver nitrate or potassium permanganate of the excess of potassium thiocyanate used as precipitant.

3. Treatment of the precipitate with alkali, producing cuprous oxide, which, if then treated with ferric chloride and hydrochloric acid, produces ferrous chloride, which may be determined.

4. Similar conversion to cuprous oxide, treatment of this with hydrochloric acid, and titration of the resulting cuprous chloride solution by potassium permanganate.

5. Direct titrimetric determination of the cuprous thiocyanate by potassium permanganate or potassium iodate.

In all these determinations it is necessary to separate the precipitated cuprous thiocyanate from the excess of potassium thiocyanate used as precipitant. If, however, the thiocyanate is allowed to react with ferric sulphate it produces an equivalent amount of ferrous sulphate. The solution now contains Cu^{++}, Fe^{++}, Fe^{+++}, K^+, H^+, SO_4^- and undissociated ferric thiocyanate complex. Interference of CNS^- can be prevented by addition of mercuric salts, which form stable complexes, soluble in excess of mercuric salt solution. An amount of mercuric salt sufficient to decolorise the solution completely will prevent interference. Diphenylamine may then be used as an internal indicator for a titration, using potassium dichromate, of the iron-II equivalent to the copper.

Oxidising agents, in particular nitrous fumes, must be absent. Nitric acid should therefore be removed by a preliminary evaporation to white fumes with sulphuric acid. Reduction of the copper solution is best achieved by sulphur dioxide in neutral or slightly acid solution. Excess of sulphur dioxide can readily be removed by boiling in a current of carbon dioxide.

Precipitation of the cuprous thiocyanate must be slow. Conversion of the iron-III to iron-II must be carried out in the cold, and the acid concentration is important. The best conditions, 24 ml. sulphuric acid (1 : 4) in 200–250 ml. solution, are fortunately also the appropriate conditions for the subsequent titration with dichromate. A large excess of potassium thiocyanate must be avoided, since it will tend to cause reversal of the ferric-ferrous change, and an excessive amount of ferric sulphate will be necessary.

Catalytic oxidation of iron-II by the atmosphere in the presence of copper-II ions is prevented by the carbon dioxide atmosphere which remains after removal of the sulphur dioxide.

16

The presence of appreciable amounts of chloride interferes, owing to the formation of mercuric chloride, which is an unsuitable mercury salt for the masking process.*[1]

The error in the determination is always less than 0·4 per cent.

The method, in a modified form, has been applied to the determination of copper in brasses or bronzes, where it proves both rapid and accurate.[2]

Procedures

Titrimetric determination of copper. Fit a 500-ml. conical flask with entry and exit tubes. It is convenient if the entry tube is fitted with a three-way tap to simplify the change-over from sulphur dioxide to carbon dioxide. In the flask place the neutral or slightly acid solution, free from oxidising agents, and containing about 0·15 g. copper as sulphate. Dilute to 200–250 ml., and saturate with sulphur dioxide in the cold. Continuing to pass sulphur dioxide, bring just to boiling, and add 0·2 N potassium thiocyanate solution drop by drop from a burette, shaking continuously, until precipitation is complete. Note the volume of precipitant used. The precipitate should be completely white, and the supernatant liquid should be colourless. A permanent greenish-yellow colour indicates incomplete removal of oxidising agents, or too high an acidity.

When precipitation is complete pass a rapid stream of carbon dioxide, washed with potassium permanganate, through the solution until the issuing gas no longer decolorises a dilute acid solution of permanganate. Allow the solution to cool to room temperature, still passing carbon dioxide. Add to the solution 25 ml. sulphuric acid (1 : 4). Add drop by drop, and shaking vigorously, a 0·5 M solution of ferric sulphate until all the precipitate has dissolved, giving a solution strongly coloured with ferric thiocyanate.

Add in one portion a volume of 0·2 M mercuric nitrate or mercuric thiocyanate solution equal to the volume of potassium thiocyanate used as precipitant. The red colour is replaced by the mixed colour of copper-II and excess iron-III. Add 10 ml. phosphoric acid (sp. gr. 1·75), which leaves only the blue of copper-II. Add 3 drops 1 per cent solution of diphenylamine in concentrated sulphuric acid, and titrate immediately with 0·1 N potassium dichromate solution.

1 ml. 0·1 N potassium dichromate≡6·357 mg. Cu.

---

* It may be noted that the mercurous nitrate method (p. 185) is free from many of the limiting factors of this method.

Determination of copper in brasses or bronzes. Dissolve the alloy, and separate tin and lead by any conventional method. Neutralise the residual sulphuric acid solution, which is free from nitrates, with ammonium hydroxide. Add 10 ml. sulphuric acid (1 : 5), dilute to 300 ml., and add 25 ml. 5 per cent sodium or ammonium bisulphite solution. Warm nearly to boiling, and add sufficient N potassium thiocyanate solution to ensure complete precipitation and leave a slight excess. Cool the solution and set it aside for 2 hours. Filter through paper, and wash the precipitate with 0·1 per cent potassium thiocyanate solution until all sulphur dioxide has been removed.

Transfer the copper thiocyanate precipitate and the filter to a conical flask, and treat in the cold with sufficient 0·5 M ferric sulphate in 0·5 M sulphuric acid to dissolve the precipitate with stirring. Dilute to about 150 ml., add at least 1 ml. M mercuric sulphate solution for every 60 mg. copper and 10 ml. sulphuric-phosphoric acid mixture (30 ml. concentrated sulphuric acid and 50 ml. phosphoric acid, sp. gr. 1·7, diluted to 100 ml. with water). Add 2 drops 0·3 per cent aqueous barium diphenylamine sulphonate solution, and titrate with 0·1 N potassium dichromate solution in the usual way.

Determine the remaining elements in the filtrate from the thiocyanate precipitate by normal methods, after removal of sulphur dioxide by boiling and destruction of thiocyanate with bromine.

[1] F. Burriel-Marti and F. Lucena-Conde, *Anal. Chim. Acta*, 1948, **2**, 230.
[2] *Idem, Anal. Fis. Quím.*, 1951, B, **47**, 459.

(c) BY OXIME PRECIPITANTS

Since bromate oxidises hydroxylamine to the nitrate stage, the reaction is more sensitive than that between the ferric ion and hydroxylamine, where only the nitrous oxide stage is reached. Moreover, when certain types of oximes are treated with bromate, oxidation or bromination of the aldehyde, formed by decomposition of the precipitate, takes place, with resulting gain in sensitivity.[1]

In certain cases, such as the copper salicylaldoxime complex, no simple equation can be written for the reaction, which must be regarded as empirical, but the oxidation may be used as a method for determining the metal.

The reactions with hydroxylamine produced by decomposition of the oxime are as follows:

$$BrO_3^- + NH_2OH \longrightarrow NO_3^- + Br^- + H_2O + H^+.$$
$$BrO_3^- + 5Br^- + 6H^+ \longrightarrow 3Br_2 + 3H_2O.$$

Because of the great sensitivity of the reaction it is particularly suited to the determination of small amounts. In any case the amount of hydroxylamine present should not exceed 20 mg.

Either salicylaldoxime or α-benzoinoxime may be used as a precipitant for copper. In the case of the former, if the hydrochloric acid solution is boiled before adding the bromate further consumption of bromate occurs, so that defined procedure should be employed. Results with the copper-α-benzoinoxime complex are slightly high, possibly because it is difficult to wash the precipitate completely free from excess reagent.

Nickel may be determined in a similar fashion to copper, using the dimethylglyoxime complex.

Procedures

Titrimetric determination of copper. Precipitate the copper by salicylaldoxime or α-benzoinoxime according to established conditions, filter on an X4 sintered-glass filter crucible, and wash the precipitate. Dissolve in 25 ml. concentrated hydrochloric acid, and draw the solution into the flask in which the titration is to be carried out. Wash the crucible with 25 ml. water. If necessary add more acid and water, keeping a 1 : 1 ratio.

Fit the flask with a two-holed rubber stopper carrying a dropping funnel and a glass stopcock. Add sufficient 0·1 N potassium bromate to give at least 10 ml. excess. If necessary add water to bring the acid concentration to 3–4 N. Mix, and set aside for 15 minutes after closing the flask.

Add to the funnel a measured excess of 0·1 N arsenious oxide solution containing 1–2 drops methyl orange indicator. Cool under the tap, and allow the bulk of the arsenious oxide solution to be drawn into the flask. When the indicator colour persists after shaking, open the stopcock and allow the remainder of the arsenious oxide solution and washings to run in from the funnel.

Remove the stopper, and titrate the excess arsenious oxide with 0·1 N bromate, using more indicator if necessary. The most critical part of the operation is the prevention of the escape of bromine when adding arsenious oxide solution to the closed system.

Salicylaldoxime: 1 ml. 0·1 N potassium bromate≡0·4541 mg. Cu.

α-Benzoinoxime: 1 ml. 0·1 N potassium bromate≡1·0595 mg. Cu.

Titrimetric determination of nickel. Precipitate the nickel with dimethylglyoxime by a normal procedure, filter off the precipitate, and treat it as described for the copper precipitate.

1 ml. 0·1 N potassium bromate≡0·2445 mg. Ni.

[1] N. H. Furman and J. F. Flagg, *Ind. Eng. Chem. Anal.*, 1940, **12**, 738.

10. TITRIMETRIC DETERMINATION OF FLUORIDE

The determination of fluoride ion by titration with aluminium chloride depends on the formation in nearly neutral solution of the stable complex ion AlF_6^{3-}, the sodium salt of which is but slightly soluble in water containing an excess of sodium ion or in ethanolic solution. Previous methods based on this reaction utilise acid-base indicators to mark the end-point, because of the sharp change in pH due to hydrolysis of the titrant. Electrometric methods have also been applied.

The end-point may be indicated[1] by the production of a highly coloured lake with the dye Eriochromcyanin R (which has been used extensively for the colorimetric determination of aluminium). The end-point is not sharp if the temperature falls below 85°–90°C. It is found convenient to use very gently boiling solutions, or solutions at temperatures just below boiling. Vigorous boiling causes precipitation of the lake formed at the end-point. Similar results are obtained both with freshly prepared and with aged solutions of Eriochromcyanin R.

The standard aluminium chloride solution is found to decrease in normality by about 0·6 per cent over a period of 2 months.

Many bivalent and tervalent ions interfere by giving coloured products, and sulphate ion interferes with the formation of the lake. However, interference from chromium, lead, nickel, carbonate, silicate, sulphate and sulphide, when present in appreciable amounts, can be overcome by distilling out the fluoride as fluosilicic acid according to the method of Willard and Winter.[2]

The method has also been used for the determination of fluorine in organic compounds after decomposition of the compounds with potassium in a nickel bomb.[3]

Procedures

Standardisation of aluminium chloride. Add 2 drops phenolphthalein indicator to 20 ml. standard sodium fluoride solution, and adjust the pH with micro-drops of 0·1 N sodium hydroxide solution and 0·1 N hydrochloric acid until the pink colour just disappears. Add 10 g. pure sodium chloride (sulphate-free) and 4 drops 0·1 per cent aqueous Eriochromcyanin R indicator solution. The solution should be yellow. Heat just to boiling, and adjust the pH again if necessary, *i.e.*, until the colour is just yellow. The solution should be saturated with sodium chloride.

Add 0·25 N aluminium chloride solution slowly from a microburette to the hot solution. Maintain the temperature just below boiling, and do not add the titrant more rapidly than 1 drop every 2–3 seconds. Near the end-point add the titrant even more slowly, and mix well. Just prior to the end-point the solution darkens, and then changes sharply to pink at the end-point. The indicator blank is 0·002–0·004 ml.

Determination of fluoride after distillation as fluosilicic acid. Titrate aliquots of the distillate, using 2 drops indicator and 5 g. sodium chloride for each 10 ml. solution. The indicator blank is 0·015 ml. for a 50-ml. aliquot, or 0·03 ml. for a 100-ml. aliquot.

1 ml. 0·25 N aluminium chloride≡4·75 mg. F.

[1] J. H. Saylor and M. E. Larkin, *Analyt. Chem.*, 1948, **20**, 194.
[2] H. H. Willard and O. B. Winter, *Ind. Eng. Chem. Anal.*, 1933, **5**, 7.
[3] M. J. Gilbert and J. H. Saylor, *Analyt. Chem.*, 1950, **22**, 196.

11. TITRIMETRIC DETERMINATION OF GERMANIUM

(a) IODOMETRIC

Potassium thiogermanate is formed when germanium solutions buffered with potassium acetate are treated with hydrogen sulphide or with potassium sulphide solution:[1]

$$2Ge^{++++} + 5S^{=} \longrightarrow Ge_2S_5^{=}.$$

Treatment with standard iodine solution then decomposes the thiogermanate, liberating sulphur, and the excess iodine may be determined by titration with standard thiosulphate solution:[2]

$$Ge_2S_5^= + 5I_2 \longrightarrow 5S + 2Ge^{++++} + 10I^-.$$

For quantitative formation of the thiogermanate a high concentration of sulphide ion is necessary. The sulphide is therefore best added as a solution of potassium hydroxide saturated with hydrogen sulphide. Carbon dioxide removes excess sulphide ion from the solution after formation of the thiogermanate, without decomposing the latter.

Metals giving insoluble sulphides in acid solution (pH 4·6) interfere. High results are obtained when sodium chloride is present throughout the sulphiding process, and this is thought possibly to be owing to formation of a higher thiogermanate, $Ge_2S_7^=$. On the other hand, low results, possibly owing to adsorption of thiogermanate on the liberated sulphur, result if sodium chloride is added after the sulphiding process and immediately before the titration.

Solution required

Potassium sulphide. Dissolve 8 g. potassium hydroxide in 100 ml. water, cool to 0° C. to avoid formation of thiosulphate or polysulphide, and saturate with hydrogen sulphide.

Procedure

Titrimetric determination of germanium. Place a measured amount of the germanium solution in a 10-inch test-tube, and dilute to 25 ml. Add 20 ml. potassium sulphide solution and then pour 15 ml. 2·5 M acetic acid carefully down the side of the test-tube, avoiding violent effervescence at the surface of the liquid. Set aside for 5 minutes. Bubble carbon dioxide rapidly through the solution for 20 minutes, stirring mechanically to promote removal of the excess hydrogen sulphide. Transfer the solution to a large beaker, dilute to 1 litre with water, and add a measured excess of standard iodine solution. Allow to stand for 15 minutes, and back titrate with standard thiosulphate solution, using starch indicator.

1 ml. 0·1 N iodine ≡ 1·452 mg. Ge.

---

[1] H. H. Willard and C. W. Zuehlke, *J. Amer. Chem. Soc.*, 1943, **65**, 1887.
[2] *Idem, Ind. Eng. Chem. Anal.*, 1944, **16**, 322.

(b) ALKALIMETRIC

The majority of published procedures for the determination of germanium suffer from lack of selectivity, and, in particular, from interference from arsenic, with which germanium is often associated. The most widely used procedure is the tannin method, with which a separation from all other metals apart from niobium, tantalum and tungsten is said to be theoretically possible. It is not, however, an attractive method since at low concentrations of germanium the precipitate tends to become colloidal, and a tedious preliminary ignition at 600° C., followed by repeated nitric acid oxidation, is necessary to ensure the absence of volatile germanous oxide before the final ignition at 900° C.

Cluley[1] has recently developed a titrimetric method based on an earlier method due to Tchakirian,[2] in which an aqueous solution of germanium dioxide is allowed to react with mannitol to form a strong complex acid. The mannitogermanic acid thus formed can be titrated alkalimetrically or iodometrically. Tchakirian stated that the method was not applicable in the presence of strong acids. Cluley has examined three different procedures based on methods used for the determination of boron, using germanium solutions containing free sulphuric acid. These are as follows:

1. The solution is neutralised with calcium carbonate and mannitol added. It is then titrated with alkali to the phenolphthalein end-point.[3]

2. The solution is adjusted with alkali to the p-nitrophenol end-point. After addition of mannitol the solution is then titrated to the phenolphthalein end-point.[4]

3. In a modification of Method 2 the pH before addition of mannitol and that at the end of the titration are the same.[5, 6]

All three methods are found to be satisfactory for 1–10 mg. of germanium, titrating against 0·02 N sodium hydroxide solution. Methods 2 and 3 are more rapid, since filtration is avoided. In Method 2 the net pH change is from 6·0 to 8·4, hence in the presence of substances exerting a buffering action a greater amount of alkali is consumed. In Method 3, where the net pH change is zero, buffering action has no effect.

The neutralisation of the mannitogermanic acid is obviously incomplete, and an empirical relationship exists between the amounts of germanium and alkali solution. It is necessary, therefore, to establish this relationship by titration of known amounts of germanium under standard conditions. Method 3 was preferred since it appears to be less susceptible to errors caused by other ions.

The results obtained in a series of standardisation titrations are shown in Table X.

TABLE X

Germanium present, mg.	Weight of germanium equivalent to 1 ml. 0·0185 N NaOH, mg.
1	1·461
3	1·367
6	1·366
10	1·359
20	1·371

For complete neutralisation of mannitogermanic acid the theoretical relationship is 1 ml. 0·0185 N sodium hydroxide solution per 1·343 mg. germanium. In practice the germanium equivalence of the titrant will vary slightly with the weight of germanium titrated, but for the small amounts involved the use of a mean equivalence factor will occasion little error, since the divergence from the theoretical relationship is only slight.

Precipitation as sulphide separates germanium from boron and many of the bases otherwise likely to be precipitated by alkali during the course of the titration. Arsenic, with which germanium is commonly associated, would precipitate under the same conditions, hence its effect was studied in some detail.

As much as 100 mg. arsenic-III has no influence on the titration of 10 mg. or less of germanium, but similar amounts of arsenic-V interfere because of the buffering effect. The difficulty is overcome by reducing to the tervalent stage with sulphur dioxide before titrating.

Traces of antimony or tin can be tolerated, but large amounts interfere because of the formation of insoluble hydroxides or basic salts.

The presence of germanium in the reagents is unlikely, but blank determinations are necessary because of the possibility of deriving traces of boron from reagents or glassware. All glassware should be made from boron-free glass where strongly alkaline solutions are used.

The titrations may be carried out using a pH meter if desired, but it is helpful to retain the use of the indicator so that readings need only be taken when the indicator colour shows an approach to pH 6·2.

The method is equally satisfactory for large amounts of germanium, for which a stronger solution of sodium hydroxide should be used.

Solutions required

Sodium germanate. Transfer 1·4408 g. pure ignited germanium dioxide to a platinum crucible, fuse with 5 g. sodium carbonate, and dissolve the cold melt in hot water. Just acidify the solution with dilute sulphuric acid, boil to eliminate carbon dioxide, cool and dilute to 1000 ml.

$$1 \text{ ml. solution} \equiv 1 \text{ mg. Ge.}$$

Buffer solution. Mix 33·9 ml. 0·1 M citric acid with 66·1 ml. 0·2 M disodium hydrogen phosphate. pH=6·2.

Standard sodium hydroxide solution. Transfer suitable volumes of the standard germanate solution, covering the range 1–20 mg. germanium, to 250-ml. flasks. Dilute each volume to about 50 ml., boil for 5 minutes to eliminate any carbon dioxide, and cool.

Add 7 drops bromocresol purple indicator solution, and add approximately 0·02 N carbonate-free sodium hydroxide solution until a value of pH 6·2 is reached as indicated by colour comparison with an equal volume of pH 6·2 buffer solution containing the same amount of indicator solution. Add 10 g. mannitol and titrate with the alkali until pH 6·2 is again reached. Carry out a blank and correct the titration figures accordingly.

From the corrected titration figures calculate the equivalence of the alkali solution in terms of mg. germanium per ml. for each weight of germanium titrated. The germanium equivalence will vary slightly with the germanium content, but it may be permissible to use a mean equivalence factor.

Procedure

Titrimetric determination of germanium. Adjust the sample solution, free from large amounts of antimony or tin, and from acids other than sulphuric acid, so that the volume is about 100 ml. and the germanium content is 1–20 mg. Transfer the solution to a conical flask and add sufficient sulphuric acid (1 : 1) to produce a concentration of this acid of 5·5 N. Pass a rapid stream of hydrogen sulphide through the solution for 30 minutes, stopper the flask, and set aside overnight.

Filter through a fine filter paper, and wash the precipitate well with 5·5 N sulphuric acid saturated with hydrogen sulphide. Dissolve the precipitate through the paper into a boron-free flask or platinum dish with three successive portions of diluted ammonia solution (2 : 1). Wash the paper first with dilute ammonia solution (1 : 9) and finally with hot water. Add 20 ml. 6 per cent w/v hydrogen peroxide and set aside for 10 minutes in the cold to ensure complete oxidation of the sulphides.

Add 5 ml. freshly prepared 5 N sodium hydroxide solution and boil vigorously until all ammonia and oxygen have been evolved, adding hot water from time to time to maintain the volume of the solution constant. Add 8 ml. dilute sulphuric acid (1 : 6) and boil for 10 minutes to remove any oxides of nitrogen formed from the oxidation of the ammonia. If desired, the solution may now be transferred to ordinary glassware.

Pass a rapid stream of sulphur dioxide into the hot solution until it is cold (about 30 minutes). Boil off excess of sulphur dioxide, ensuring its complete removal by boiling for 5 minutes after the smell of the gas can no longer be detected. This reduction may be omitted if it is known that arsenic is absent.

Add 2 drops bromocresol purple indicator, and then add freshly prepared 5 N sodium hydroxide solution until the solution is just alkaline. Add N sulphuric acid until the indicator just turns yellow, dilute to about 80 ml., boil for 5 minutes to eliminate any carbon dioxide, and cool.

Add 5 further drops of bromocresol purple indicator, adjust the solution to pH 6·2, add 10 g. mannitol, and titrate to pH 6·2 as in the standardisation titrations. Carry out a blank determination, correct the sample titration figure, and calculate the germanium content of the solution from the corrected figure.

[1] H. J. Cluley, *Analyst*, 1951, **76**, 517.
[2] A. Tchakirian, *Compt. rend.*, 1928, **187**, 229.
[3] E. T. Wherry, *J. Amer. Chem. Soc.*, 1908, **30**, 1687.
[4] W. F. Hillebrand, G. E. F. Lundell, H. A. Bright and J. J. Hoffman, *Applied Inorganic Analysis*, New York and London, 1953, 753.
[5] F. J. Foote, *Ind. Eng. Chem. Anal.*, 1932, **4**, 39.
[6] M. Hollander and W. Riemann, *ibid.*, 1946, **18**, 788.

12. TITRIMETRIC DETERMINATION OF IRON-III

(a) By Silver Reductor

The silver reductor was first introduced for oxidimetric titrations in 1916,[1] and was subsequently applied to the determination of iron,[2] the ferrous solution being titrated with ceric sulphate or with ammonium sulphatocerate, $(NH_4)_4Ce(SO_4)_4 . 2H_2O$.

Henry and Gelbach[3] have shown that it is not essential to use ceric solutions for the final titration, and that this may be carried out with dichromate, using diphenylamine sulphonic acid as indicator, when a sharp end-point is obtained. The most satisfactory acidity is the range 0·5–1·5 N in hydrochloric acid. Large amounts of vanadium (more than 100 mg. in 200 ml. solution) produce an indistinct end-point, but moderate amounts of chromium, manganese, titanium and vanadium do not interfere.

Procedure

Titrimetric determination of iron-III. To a 25-ml. sample of ferric solution add 5 ml. concentrated hydrochloric acid and dilute to 50 ml. with water, achieving a final acidity of about N hydrochloric acid. Pass the solution through a silver reductor at about 30 ml. per minute, and wash through with 150 ml. N hydrochloric acid in small portions. Add 5 ml. syrupy phosphoric acid and 5–6 drops indicator, and titrate with standard dichromate solution.

$$1 \text{ ml. } 0 \cdot 1 \text{ N potassium dichromate} \equiv 5 \cdot 584 \text{ mg. Fe.}$$

[1] G. Edgar, *J. Amer. Chem. Soc.*, 1916, **38**, 297.
[2] G. H. Walden, L. P. Hammett and S. M. Edmonds, *ibid.*, 1934, **56**, 59.
[3] J. L. Henry and R. W. Gelbach, *Ind. Eng. Chem. Anal.*, 1944, **16**, 49.

(b) By Reduction With Sulphurous Acid

The reducing action of sulphurous acid on ferric salts has not been fully studied. Some authorities indicate that the reaction proceeds best in a neutral solution, being incomplete in a sulphuric acid or a hydrochloric acid medium. Others consider that a slightly acid medium gives the best results. The amount of iron, it is claimed, must not be too great.

The equation for the reaction is usually stated to be as follows:

$$Fe_2(SO_4)_3 + SO_2 + H_2O \longrightarrow 2FeSO_4 + H_2SO_4.$$

The mechanism of the reduction, however, is undoubtedly more complicated than this equation suggests. A red compound is first formed which appears to be a salt of ferrisulphurous acid, with the possible formula $Fe[Fe(SO_3)_3]$. This breaks down thus:

$$Fe[Fe(SO_3)_3] \longrightarrow FeS_2O_6 + FeSO_3,$$

and hydrolysis then produces ferrous sulphate and sulphurous acid:

$$FeS_2O_6 + H_2O \longrightarrow FeSO_4 + H_2SO_3.$$

An investigation of the experimental conditions has led to the following conclusions:[1]

1. An approximately neutral medium gives the best results for quantitative reduction (using $c.$ 0.07 g. iron-III). An increase in acidity increases the time required. For example, with a solution which was 0.1 N to sulphuric acid reduction was complete after 15 minutes' boiling. When the acidity was increased to 0.2 N, 45 minutes were required. With a solution which was N to sulphuric acid 150 minutes were required.

2. Passage of sulphur dioxide through the cold solution with subsequent boiling is more efficient than passing the gas through the boiling solution.

3. As the amount of iron-II increases the reaction is slowed down by the ordinary mass action effect.

4. The reaction is accelerated by the addition of potassium thiocyanate. In its presence the reduction can be effected rapidly, even in solutions of fairly high acidity. The mechanism of the reaction is suggested to be the partial replacement of $SO_3^=$ by CNS^- in the intermediate $Fe[Fe(SO_3)_3]$. Since the complex is only formed when sulphur dioxide is passed through the cold solution, and not when it is passed through the boiling solution, it is assumed that this accounts for the increased rate of reduction in the cold. This appears more probable than the usually accepted explanation that the increased efficiency is due to the increased solubility of sulphur dioxide in the cold solution.

The thiocyanate-accelerated reaction has possible applications in determinations where it is inconvenient to work in a neutral solution because of hydrolysis of other components such as titanium. At an acidity of 1·0–1·5 N in sulphuric acid titanium is not precipitated, even after prolonged boiling.

Satisfactory results are obtained when determination of iron-III is carried out in the presence of appreciable quantities of titanium.

Procedure

Titrimetric determination of iron-III. Add 10 ml. 0·1 N potassium thiocyanate solution to the solution of iron-III, which should be less than 2 N in sulphuric acid. Saturate the solution with sulphur dioxide in the cold, or add 50 ml. sulphurous acid solution. Heat slowly to boiling. The solution rapidly becomes colourless or only slightly yellow, and the reduction is then complete. Pass carbon dioxide until all the sulphur dioxide has been removed, cool the solution, and add 10 ml. 0·1 M mercuric nitrate or mercuric sulphate solution.

Titrate in the usual way using potassium dichromate solution or potassium permanganate solution, with barium diphenylamine sulphonate as indicator.

1 ml. 0·1 N potassium dichromate≡5·584 mg. Fe.

[1] F. Burriel-Marti and F. Lucena-Conde, *Anal. Chim. Acta*, 1949, 3, 547.

(c) By Stannous Chloride Reduction

When stannous chloride is used to reduce iron-III before titration with an oxidimetric agent, the excess is normally removed by the addition of mercuric chloride. Unless great care is taken in limiting the excess of mercuric chloride the determination may be vitiated by the formation of grey colloidal mercury.

Silicomolybdic acid may be used instead of mercuric chloride.[1] It is reduced to molybdenum blue by the excess of stannous chloride. On adding standard dichromate solution the blue colour in turn disappears at the point where all the excess stannous chloride has been oxidised, and the oxidation of the iron-II just begins. Further addition of dichromate oxidises the iron-II, and the end-point is observed using N-phenylanthranilic acid. Consequently, the volume of dichromate consumed from the point at which the molybdenum blue disappears to the point at which the red colour of the N-phenylanthranilic acid appears is equivalent to the iron being determined.

At the beginning of the determination 7·5 ml. concentrated hydrochloric acid should be present. If the iron solution already contains hydrochloric acid the total amount should be adjusted accordingly.

Blank tests may be run for impurities in the reagents, but for routine purposes this need not be done. The authors state that "it would seem wise to filter out any cloudy material such as silica during the application of the method" but do not indicate whether or not they found this necessary.

Ceric sulphate may also be used for the final determination instead of potassium dichromate.[2] No experimental details are supplied, however, apart from the statement that 32 ml. sulphuric acid (1 : 1) is added to the iron solution in addition to the 8 ml. present in the added ceric sulphate solution. The authors are very obscure as to whether blank corrections should be applied or not. It would seem from the results that the method is less satisfactory than that using dichromate.

Solutions required

Silicomolybdic acid indicator solution. Mix 13·2 g. sodium silicate pentahydrate dissolved in water with a solution containing 35 g. ammonium molybdate tetrahydrate, and add 49 ml. concentrated sulphuric acid. Dilute to 2 litres.

N-phenylanthranilic acid indicator solution. Dissolve 1 g. N-phenylanthranilic acid in a little warm water containing 0·6 g. sodium carbonate. Dilute to 1 litre.

Procedure

Titrimetric determination of iron-III. Transfer 15 ml. hydrochloric acid (1 : 1, sp. gr. 1·1) to a 500-ml. conical flask, and add 25 ml. water. Place a small watch-glass on top of the flask, and boil the acid until the volume has been reduced by 5–7 ml., driving the air from above the solution past the water seal formed under the watch-glass. Transfer the flask to a hot-plate regulated so that the solution boils slowly.

Add 45 ml. of the iron solution. If it is desired to use a smaller aliquot, the volume should be adjusted to about 45 ml. When the whole solution boils add 2 N stannous chloride solution (containing less than 50 ml. hydrochloric acid per litre) until the yellow colour of the iron solution disappears. (When 45 ml. 0·1 N ferric chloride solution were treated, 2·75 ml. 2 N stannous chloride solution were required and the volume of the solution at this stage was about 80 ml.)

Add 5 drops silicomolybdic acid indicator solution. The solution should turn blue. Cool, keeping the watch-glass in position to avoid risk of oxidation. Titrate with standard potassium dichromate to the disappearance of the blue colour, which may be noted by looking through the solution on to the surface of a daylight fluorescent lamp placed at the same height and taking the first major change in shade upon swirling after the addition of a drop of dichromate solution. With small samples this change is from blue to yellow or light green, and even with larger samples, in daylight, a fairly good blue to green end-point can be detected without the lamp. Allow 30–60 seconds between addition of each drop in the final stages.

When this end-point has been reached add a mixture of 85 ml. water and 40 ml. sulphuric acid (1 : 1) which has been boiled for at least 3 minutes. Boil the whole for 20 minutes to destroy the silico-molybdic acid, otherwise a muddy brown colour will appear at the second end-point. Remove from the hot-plate and cool in running water.

Titrate with potassium dichromate solution, with constant swirling, using 8 drops N-phenylanthranilic acid indicator solution. As the indicator is destroyed rather readily it should be added in instalments. The final volume of solution at the end-point should be about 200 ml.

The amount of iron present is equivalent to the amount of dichromate solution consumed between the two end-points.

1 ml. 0·1 N potassium dichromate ≡ 5·584 mg. Fe.

[1] A. C. Titus and C. W. Still, *Ind. Eng. Chem. Anal.*, 1941, **13**, 416.
[2] *Idem, ibid.*, 1942, **14**, 212.

13. GRAVIMETRIC DETERMINATION OF LITHIUM

(a) As Lithium Aluminate

Lithium has been determined as (1) the fluoride, (2) the phosphate, and (3) the triple zinc uranyl acetate. Grothe and Savelsburg[1] have investigated these procedures, in particular from the point of view of their suitability for the rapid determination of lithium in reasonably dilute solution and in the amounts normally met with industrially. They have reported as follows:

1. Lithium fluoride is precipitated in ammoniacal solution by ammonium fluoride. An empirical correction is necessary owing to the high solubility of lithium fluoride (0·280 g. per l.). Previous authors[2] have recommended the addition of 2 mg. to the weight of precipitate observed, for each 7 ml. filtrate. This was found to be an excessive correction, an addition of 2 mg. per 10 ml. filtrate giving better results. However, in the presence of magnesium, prior precipitation of magnesium as magnesium ammonium phosphate led to some coprecipitation of lithium, while the presence of phosphate affected the subsequent precipitation of the remainder of the lithium. The method proved to be particularly poor when estimating small amounts of lithium in the presence of large amounts of magnesium. Prior removal of the magnesium leaves so much ammonium salt in the solution that it is impossible to concentrate it to the small bulk necessary for optimum precipitation of the lithium. In such cases the empirical correction might be as much as one-third of the total weight of precipitate; obviously an unsound practice.

2. Precipitation as phosphate gave variable errors, positive when the amount of lithium was large and negative when the amount of lithium was small. There also appeared to be occlusion of sodium phosphate when this salt was used to depress the solubility of the precipitate.

3. Satisfactory precipitation of the triple acetate was only possible in concentrated solution. Prior removal of magnesium was therefore not feasible, but if the magnesium were not so removed the composition of the precipitate was variable, containing both magnesium and lithium in proportions depending on the relative concentrations of these two ions in the solution. The presence of calcium also interfered.

Aluminium has been determined as lithium aluminate, the precipitate having the composition $LiH(AlO_2)_2.5H_2O$. On ignition this is converted to the form $2Li_2O.5Al_2O_3$. Conditions have been determined by Grothe and Savelsburg which are suitable for the precipitation of lithium in the same form.

The compound contains 4·88 per cent lithium so that the factor for conversion of the ignited precipitate is favourable. The solubility of the precipitate in water is 8×10^{-3} g. per l., and in the strongly alkaline medium found to be the most suitable for the precipitation

17

the solubility is 0·09 g. per l., or $1·75 \times 10^{-4}$ mole per litre. This compares very favourably with the solubility of lithium fluoride. The precipitate is flocculent, settles readily, and is easy to filter and wash if the proper conditions are observed.

For best results precipitation must be carried out by sodium aluminate solution in the cold, and the precipitate must be washed with cold water, heating at any stage resulting in a slimy precipitate which is difficult to handle. At least double the theoretical amount of sodium aluminate must be used. The lithium must be precipitated from an acid solution (pH 2·0–3·0), but the precipitating solution must be strongly alkaline. It is prepared by addition of concentrated sodium hydroxide solution to a potassium alum solution, producing a mixed solution with a pH in the range 12·5–13·1. After precipitation the pH of the supernatant liquid must also be adjusted to this value. If these conditions are not observed coprecipitation of aluminium hydroxide occurs, and this, once precipitated, cannot subsequently be redissolved. The pH of the lithium solution may readily be adjusted by using an indicator, but no indicator is sufficiently sensitive in the alkaline range of the precipitating solution, so that this and the mixed solutions after precipitation must be checked electrometrically.

Apart from these drawbacks the method is claimed to be rapid and accurate. Since the volume of lithium solution may be large (250–300 ml.) magnesium may be removed by a conventional method, though no figures are given to show the effect of the presence of either magnesium or phosphate ions. Nor is any indication given of the investigation of other interferences.

The precipitate is washed with cold water, ignited, and weighed as $2Li_2O \cdot 5Al_2O_3$.

The authors point out that the determination of aluminium in the presence of lithium may give high results owing to precipitation of lithium aluminate in ammoniacal solution, and recommend that in such cases aluminium should be determined by precipitation as phosphate from acetic acid solution.

Solution required

Alkaline sodium aluminate solution. Dissolve 50 g. potassium alum with warming in 900 ml. water. Cool, and while keeping the solution cool add a concentrated solution of 20 g. sodium hydroxide, stirring so that the precipitate which forms is redissolved. Set the solution aside overnight, filter, and adjust the pH to 12·6. Make up to 1 litre with water.

Procedure

Gravimetric determination of lithium. Adjust the lithium solution to pH 3·0, and add 40 ml. precipitating solution for each 10 mg. lithium. Adjust the mixed solutions to pH 12·6 by dropwise addition of sodium hydroxide solution. Set aside for a short time, and decant through a filter, washing the residue of the precipitate into the filter with cold water. Wash with further amounts of cold water until the washings are no longer alkaline to phenolphthalein. Ignite and weigh.

$$\text{mg. ppt.} \times 0·0488 \equiv \text{mg. Li.}$$

[1] H. Grothe and W. Savelsburg, *Z. anal. Chem.*, 1937, **110**, 81.
[2] J. T. Dobbins and J. P. Saunders, *J. Amer. Chem. Soc.*, 1932, **54**, 178.

(b) As Lithium Zinc Uranyl Acetate

Grüttner[1] has carried out a critical investigation of the determination of small amounts of lithium by precipitation as sulphate, aluminate or triple zinc uranyl acetate. The sulphate method was found to give rise to considerable errors, and the aluminate method was judged to depend too much on precise control of pH and other factors to make it satisfactory for determination of small amounts. The report on the triple acetate method was favourable.

The precipitate has the formula $LiZn(UO_2)_3(CH_3COO)_9 . 6H_2O$, giving a very favourable factor. Deposition of the precipitate is carried out in the cold by a solution of zinc uranyl acetate which is saturated with respect to the lithium triple salt. As a consequence it is essential that the precipitating solution, the wash solution, and the suspension before filtering should all be maintained at constant temperature. For this purpose, however, a water-bath at room temperature is sufficient to prevent objectionable fluctuations of temperature, and thermostatic control is not required.

It is recommended that each new batch of precipitating solution should be checked against a solution of known lithium content.

Solutions required

Standard lithium test solution. Treat a fairly concentrated solution of pure lithium chloride with ammonium carbonate solution to precipitate the lithium as carbonate. Filter this off, wash with water until chloride-free, dry, and heat in an aluminium block at 250°–280° C. for 1–2 hours. Although this preparation will not be spectroscopically pure, the error from the sodium content will be less than experimental error.

Weigh accurately the required amount of this product, dissolve it in the calculated amount of hydrochloric acid or perchloric acid, and make up to a known volume. Use this solution for check analyses.

Zinc uranyl acetate solution. Dissolve 29·2 g. zinc acetate dihydrate in 250 ml. 60 per cent acetic acid by gentle warming. Cool the solution and add 35–37 g. finely powdered uranyl acetate, $UO_2(CH_3COO)_2.2H_2O$. Stir vigorously at room temperature for 12–15 hours. Add 1–2 g. solid lithium zinc uranyl acetate, and stir for 3 hours. Set the suspension aside in a water-bath at room temperature and protected from direct light for at least 4–5 days.

Immediately before use shake vigorously and filter from the solid residue, which consists of excess of uranyl acetate and lithium zinc uranyl acetate, and traces of sodium zinc uranyl acetate derived from sodium impurities in the reagents.

Wash solution. Saturate 95 per cent ethanol with solid lithium zinc uranyl acetate.

Procedure

Gravimetric determination of lithium. Evaporate to dryness the lithium solution containing 0·7–2·3 mg. lithium as chloride or perchlorate. Cool the residue and add directly to it 10 ml. filtered zinc uranyl acetate solution. The residue first dissolves and then a finely crystalline heavy bright yellow precipitate begins to form. Hasten the formation of the precipitate by stirring for 15 minutes, and then set aside in a water-bath at room temperature for 45 minutes. Filter through an X4 sintered-glass filter crucible, wash five times with 2-ml. portions of precipitating solution, five times with 2-ml. portions of wash solution, and finally twice with a few ml. anhydrous ether. Air-dry for 15 minutes, wipe the outside of the filter, acclimatize in the balance-case for 15 minutes, and weigh.

$$mg. ppt. \times 0·0456 \equiv mg. Li.$$

[1] B. Grüttner, *Z. anal. Chem.*, 1951, **133**, 36.

14. TITRIMETRIC DETERMINATION OF MOLYBDENUM

The method of Banks and Diehl[1] for the determination of thorium as molybdate (p. 273) may be used in reverse for the determination of molybdenum. It is useful for separating this element from certain elements from which it cannot readily be separated by other means. The method is particularly suitable for separating molybdenum from uranium in the analysis of molybdenum-uranium alloys.

The sample is normally dissolved in hydrochloric acid (1 : 1), but nitric acid is necessary in addition to promote solution of alloys containing 20 per cent or more of molybdenum. This nitric acid, which must be removed before subsequent determination of uranium, cannot, however, be fumed off immediately after dissolving the sample, since the use of sulphuric acid would later precipitate thorium sulphate and the use of perchloric acid would cause molybdic acid to precipitate.

The sample may be weighed directly, or, in the case of larger samples where the sample form demands it, it may be weighed, dissolved, and diluted to a known volume, aliquots then being taken for analysis.

Procedures

Titrimetric determination of molybdenum. Transfer the sample, containing 0·1–0·15 g. molybdenum trioxide, to a 400-ml. beaker, and dissolve in the minimum volume of hydrochloric acid (1 : 1). If the sample does not dissolve on warming, add a little concentrated nitric acid, and boil to aid solution. Add hydrogen peroxide to dissolve the black hydrated uranium dioxide and to oxidise both elements to the sexavalent state. Boil for 10 minutes to destroy the excess of hydrogen peroxide, and dilute to 200 ml.

Add 16 ml. glacial acetic acid and sufficient ammonium acetate to react with the mineral acid (usually 1 g.). Add 15 ml. of a slurry of filter pulp. Add slowly, with stirring, a 25 per cent excess of thorium perchlorate solution. Heat the contents of the beaker to boiling, and filter hot through an 11-cm. Whatman No. 42 filter paper. Wash the precipitate five or six times with hot dilute acetic acid (1 : 100). It is not necessary to remove the last traces of precipitate from the beaker. Two or three rinses of the beaker with wash solution are sufficient.

Transfer the filter paper and precipitate to the beaker in which the precipitation was carried out, and add 25 ml. concentrated hydrochloric acid. Stir the contents of the beaker until the paper is disintegrated, add 75 ml. water, and heat just to boiling. Prolonged boiling must be avoided, since molybdenum is reduced and the filter paper is decomposed.

Filter while hot through a Whatman No. 42 paper into a 400-ml. beaker and wash the filter and the paper pulp five or six times with hot dilute hydrochloric acid (1 : 100).

Cool the filtrate to room temperature and pass through a Jones reductor containing amalgamated zinc into an excess of ferric alum solution (five times the theoretical amount of 10 per cent ferric alum solution) containing 2–3 ml. concentrated phosphoric acid. Titrate with 0·1 N ceric sulphate using 2 drops 0·0025 M 1 : 10-phenanthroline-ferrous sulphate as indicator. The end-point is taken as that point at which the pink colour of the solution changes to colourless or light blue.

1 ml. 0·1 N ceric sulphate≡9·595 mg. Mo.

Titrimetric determination of uranium. Concentrate the filtrate from the thorium molybdate precipitate to a convenient volume. If nitric acid was used in dissolving the sample, fume with perchloric acid. Reduce the solution in a Jones reductor, aerate, and titrate with 0·1 N ceric sulphate in the usual way.

1 ml. 0·1 N ceric sulphate≡11·904 mg. U.

[1] C. V. Banks and H. Diehl, *Ind. Eng. Chem. Anal.*, 1947, **19**, 222.

15. TITRIMETRIC DETERMINATION OF NITRATE

The determination of nitrate by reduction with ferrous sulphate is a well-known procedure. It was improved by Kolthoff and his co-workers,[1] who reduced the time of boiling required to 10 minutes by using a molybdate catalyst. An atmosphere of carbon dioxide was still considered necessary.

Leithe[2] has simplified the method by reducing the boiling time to 3 minutes and eliminating the need for a carbon dioxide atmosphere or for a catalyst. The acidity is raised to 6–8 M in sulphuric acid. The method has been applied to the analysis of fertilisers and commercial nitrates, spent mixed acids, and meat pickle salts.

Normally 0·1 N or 0·2 N solutions are used but in some cases 0·5 N solutions may be used, depending on the amount of sample taken.

Solutions required

0·2 N ferrous sulphate. Dissolve 55 g. ferrous sulphate and 20 g. sodium chloride in 100 ml. water containing a few drops of dilute sulphuric acid in a 1000-ml. graduated flask. Make up to the mark with sulphuric acid (1 : 1). The solution is stable over 1 day, and frequently for longer periods.

Procedure

Titrimetric determination of nitrate. Transfer 25 ml. sample solution, containing 25–80 mg. nitrate, to a 250-ml. conical flask. Add 25 ml. 0·2 N ferrous sulphate solution and 20 ml. concentrated sulphuric acid, whilst swirling the flask. Heat to boiling and boil for 3 minutes over a moderate flame.

Cool under the tap, add 50 ml. cold water and 1 ml. 0·0025 M 1 : 10-phenanthroline-ferrous sulphate indicator, and titrate with 0·1 N potassium dichromate solution until the colour changes from orange through brown to green.

Carry out a blank test in exactly the same way, but using 25 ml. water instead of the sample solution. Correct for the blank titration figure.

Potassium permanganate solution may be used instead of potassium dichromate solution if the sample and blank titrations are carried out in exactly the same way.

$$1 \text{ ml. } 0·1 \text{ N potassium dichromate} \equiv 2·067 \text{ mg. } NO_3^-.$$
$$\equiv 0·4470 \text{ mg. N.}$$

[1] I. M. Kolthoff, E. B. Sandell and B. Moskovitz, *J. Amer. Chem. Soc.*, 1933, **55**, 1454.
[2] W. Leithe, *Analyt. Chem.*, 1948, **20**, 1082.

16. STANDARDISATION OF OXIDISING AGENTS

The ferrous salts which are normally used for the titration of oxidising agents cannot be utilised as primary standards, either because of difficulty in obtaining the salt in pure condition, or because of ease of oxidation. While this is not serious when dichromate is being employed as oxidant, neither potassium permanganate nor cerate solutions can be prepared without further titration against some independent standard such as arsenite or oxalate.

Ferrous ethylenediamine sulphate (FES)[1] and ferrous propylene-diamine sulphate (FPS)[2] have been investigated as possible primary ferrous standards, and both of these have proved quite satisfactory. Both compounds conform to the requirements laid down by Dodge[3] for a primary standard, which are as follows:

1. The substance should be readily obtained pure.

2. It should not be affected by air at ordinary or moderately elevated temperatures.

3. It should be readily soluble in water or in ethanol.

4. It should have a high equivalent weight.

5. It should not produce any interfering substance on titration.

6. It should not be sufficiently coloured, either before or after titration, to interfere with indicators.

Examination of a wide range of ferrous diamine sulphates[2] has brought to light none, other than FES and FPS, which have the desired properties.

Ferrous ethylenediamine sulphate, $(CH_2.NH_2)_2H_2SO_4.FeSO_4.4H_2O$, equivalent weight 382, is an almost white crystalline, non-deliquescent, non-efflorescent solid with a pale green tinge. It is easy to prepare, purify and dry. It reacts stoicheiometrically with the usual titrimetric oxidising agents. After 150 days' exposure to a laboratory atmosphere it gave negative tests for ferric iron, and showed no change in composition.

Ferrous propylenediamine sulphate, $(CH_3.CHNH_2.CH_2.NH_2)_2.H_2SO_4.FeSO_4.4H_2O$, equivalent weight 396, is similar in appearance and behaviour to FES. After 50 days, loosely covered in the laboratory, or after 12 hours at 50°C., its composition was unchanged.

Solutions of both FES and FPS gradually deteriorate by about 2 per cent after 14 days, indicating that the ferrous ion is not stabilised to any extent in solution, since this amount of deterioration is comparable with that of the more usual ferrous solutions.

Procedure

Standardisation of a 0·05 N oxidant. Dissolve 0·4–0·5 g. ferrous diamine sulphate in 150 ml. water and add 10 ml. 18 N sulphuric acid. Titrate with cerate solution using Erioglaucine A or *o*-phenan-throline-ferrous sulphate as indicator, or against permanganate without an indicator. For titration against potassium dichromate add 5 ml. phosphoric acid (1 : 1) and use diphenylamine or diphenyl-amine sulphonic acid as indicator.

Synthesis of ferrous diamine sulphates

The method used follows that of Grossmann and Schück.[4] Add 60 ml. 6 N sulphuric acid to 10 g. 98 per cent solution of ethylene-diamine or 10 g. propylenediamine, and follow this with 46·3 g. (for FES) or 37·6 g. (for FPS) of ferrous sulphate heptahydrate which has been purified by dissolving in distilled water and precipitation with ethanol, followed by filtration and air-drying. Adjust the volume to 300 ml. with water, and add 300 ml. ethanol slowly with constant stirring. Filter on a Büchner funnel, wash the precipitate with 50 per cent ethanol, and redissolve in slightly acidulated water. Add two-thirds the volume of ethanol. Filter once more on a Büchner funnel, wash first with 65 per cent ethanol and then with 95 per cent ethanol. Air-dry, or dry overnight in an oven at 50°C.

If, as may happen, the compound separates as a viscous liquid, decant off the supernatant liquid and add 50 ml. acetone with constant stirring, to cause crystallisation.

Yield for FES=84 per cent: for FPS=75 per cent.

[1] K. P. Caraway and R. E. Oesper, *J. Chem. Educ.*, 1947, **24**, 235.
[2] A. J. Nutten, *Anal. Chim. Acta*, 1949, **3**, 433.
[3] F. D. Dodge, *Ind. Eng. Chem.*, 1915, **7**, 29.
[4] H. Grossmann and B. Schück, *Z. anorg. Chem.*, 1906, **50**, 26.

17. GRAVIMETRIC DETERMINATION OF PHOSPHORUS

In the gravimetric determination of phosphorus as ammonium phosphomolybdate an empirical factor for phosphorus must always be used, since although under the same conditions the precipitate has a constant composition, this does not correspond to a stoicheiometric value.

Previous methods have used ammonium sodium hydrogen phos-phate as the basis for the determination of the factor. Potassium dihydrogen phosphate has now been recommended,[1] and gives a value of the factor rather smaller than that usually employed.[2] The potassium dihydrogen phosphate is purified by repeated recrystallisation from water, and dried for 3 hours at 110°C.

Solutions required

Sulphate-molybdate reagent. Dissolve 50 g. ammonium sulphate in 500 ml. nitric acid (sp. gr. 1·36). Dissolve 150 g. powdered ammonium molybdate in 400 ml. boiling water, cool, and pour this in a thin stream, with constant stirring, into the nitric acid-sulphate solution. Dilute to 1 litre, allow to stand for 3 days, and filter into a brown glass-stoppered bottle.

Nitric-sulphuric acid. Add 420 ml. nitric acid (sp. gr. 1·40) to 580 ml. water, and to this add carefully 30 ml. sulphuric acid (sp. gr. 1·84).

Procedure

Determination of factor. Dissolve a weighed amount of potassium dihydrogen phosphate in nitric acid (1 : 1), and add 2 ml. nitric-sulphuric acid reagent. Make up to a volume of 15 ml. with water, and warm at 60°C. in a water-bath. Add rapidly 15 ml. sulphate-molybdate reagent, and set aside for 4–5 minutes. Shake the vessel for half a minute, and set aside for 2–5 hours. Filter through a sintered-glass filter tube, wash with 2 per cent aqueous ammonium nitrate solution, and then with ethanol and with acetone. Dry for 20 minutes in a vacuum desiccator at 25–60 mm., and weigh 3 minutes after removing from the desiccator.

Typical factor found:

$$\text{Wt. phosphomolybdate} \times 0\cdot01442 \equiv \text{mg. P.}$$
$$\text{Wt. phosphomolybdate} \times 0\cdot03304 \equiv \text{mg. } P_2O_5.$$

[1] Y. Tsuzuki, M. Miwa and E. Kobayashi, *Analyt. Chem.*, 1951, **23**, 1179.
[2] F. Pregl (transl. J. Grant), *Quantitative Organic Microanalysis*, 5th English Ed., London, 1951, p. 142.

18. TITRIMETRIC DETERMINATION OF PHOSPHORUS

Wilson[1] has described a new method for the determination of phosphate based on precipitation as quinoline phosphomolybdate, which is isolated and then titrated alkalimetrically. The main advantage over the older ammonium phosphomolybdate method is the greater accuracy arising from the lower solubility of the precipitate; the mean difference between duplicate determinations was 0·027, corresponding to a standard deviation of 0·024, whereas the standard deviation for the ammonium phosphomolybdate method was found to be 0·065. The method is also relatively free from interferences, possibly because of the different crystal structure of the precipitate.

Attempts to complete the determination gravimetrically were unsuccessful because of some contamination. The nature of the contaminant was not ascertained, although it was concluded that it could not be quinoline molybdate or molybdic acid since otherwise the titrimetric method would also be in error. It was not hydrochloric acid nor sodium chloride, as chloride could not be detected in the precipitate. It was thought possibly to be water obstinately retained, or sodium molybdate. The degree of contamination may be judged from the average apparent molecular weight of the precipitate, which was found to be 2225·8 against the theoretical value of 2212·8.

Precipitation as cinchonine or pyridine phosphomolybdates was also investigated, but the former caused filtration difficulties, and the molecular weight was much lower than that required by theory. Pyridine phosphomolybdate was too soluble.

Increased accuracy is obtained in the new method by weighing aliquots instead of measuring them by volume. There appears to be no reason, however, why the more convenient volumetric measurement of aliquots should not be used except when the greatest possible accuracy is required.

Possible interferences have been thoroughly investigated, and the following data have been recorded.

Lime. 1 g. lime has no effect (50 mg. P_2O_5 present).

Fluorine. 2 ml. hydrofluoric acid gives slightly high results (*e.g.*, 51·5 mg. instead of 50 mg.). This is owing to attack on the beaker, which brings silica into solution. The effect is overcome by adding boric acid. For example, 3·5 g. calcium carbonate, 5 g. boric acid and 2 ml. 40 per cent hydrofluoric acid were added to 50 mg. P_2O_5. Hydrochloric acid was then added until the calcium carbonate had dissolved, and the determination was completed as usual, when 49·94 mg. P_2O_5 were found in one determination, and 50·08 mg. when the amount of calcium carbonate was increased.

Ammonia. Ammonia interferes. For example, in the presence of 1 g. ammonium sulphate only 47·83 mg. P_2O_5 were found. Hypo-bromite is recommended as a means of destroying ammonia.

Iron. Iron has no effect. In the presence of 1 g. iron the correct result was obtained.

Magnesia. 2 g. $MgSO_4.7H_2O$ has no effect on the determination of 50 Mg. P_2O_5.

Alkali Salts. 5 g. potassium chloride or 10 g. sodium sulphate have no effect.

Citric acid and ammonium citrate. 0·5 g. of either has no effect.

Nitric acid. Substitution of an equivalent amount of nitric acid for hydrochloric acid in the determination has no effect.

Sulphuric acid. When present in an amount equivalent to the hydrochloric acid, high and erratic results are obtained. A white solid is precipitated with the quinoline phosphomolybdate. Even in the absence of quinoline and phosphate, molybdic acid is slowly precipitated when sodium molybdate is heated with dilute sulphuric acid. This does not occur with hydrochloric acid present in the concentrations used in the recommended method. Precipitation occurs much more readily when ammonium sulphate is present. The amount of $SO_4^=$ and NH_4^+ in the precipitate are small, and adsorption seems more likely than the formation of compounds.

If hydrochloric acid is present in an amount slightly in excess of the sulphuric acid, the interference is prevented, but the total amount should not be greatly in excess of 2 N. It is preferable to avoid large excesses of sulphuric acid, and the method should not be used for organic compounds which have been decomposed by the Kjeldahl method.

Solutions required

Hydrochloric acid. Concentrated acid. Dilute acid (1 : 9). 0·5 N. 0·1 N.

Sodium hydroxide. 0·5 N and 0·1 N carbonate-free solutions.

Sodium molybdate solution. Dissolve 150 g. hydrated salt in water and make up to 1 litre. This solution should not be kept too long as it dissolves silica from the glass container.

Quinoline hydrochloride. Add 20 ml. redistilled quinoline (prefer-ably synthetic) to 800 ml. hot water acidified with 25 ml. concen-trated hydrochloric acid, and stir well. Cool to room temperature, add paper pulp and again stir well. Filter through a paper-pulp pad with suction, but do not wash. Dilute to 1 litre with water.

Bromine water. Saturate distilled water with bromine at room temperature.

Perchloric acid. AnalaR, 60 per cent.

Mixed indicator solution. Mix two volumes 0·1 per cent phenolphthalein solution with three volumes 0·1 per cent thymol blue solution (both ethanolic).

Procedures

(a) *Determination of total P_2O_5 in basic slag, phosphate rock, superphosphate, etc.* (*not containing ammonia*). (*i*) Weigh 2·5 g. for samples of normal P_2O_5 content, and transfer to a 150 ml. beaker. Add 20 ml. water and 10 ml. perchloric acid.* Warm until most of the sample is dissolved, then evaporate to fuming. Continue this fuming with the beaker covered for at least 15 minutes, until the attack is complete. Allow to cool, rinse the cover with a little distilled water, add 20 ml. hydrochloric acid (1 : 9), warm carefully, and boil until all salts are in solution. Filter through a 9-cm. Whatman No. 30 filter paper in a 2-inch funnel into a 125-ml. conical flask provided with a stopper, which has been weighed to the nearest 0·5 mg., preferably with a similar flask as counterpoise. Transfer the contents of the beaker to the filter with the minimum amount of water, using a wash-bottle with a fine jet and a succession of small washes, allowing each to run through before adding the next to the filter. Wash the filter thoroughly with small portions of warm water in the same way. The total volume of filtrate and wash liquid should be about 100–110 ml. Discard the filter paper.

(*ii*) Cool the flask, stopper it, and mix the contents thoroughly. Carefully dry the outside, and weigh to the nearest 0·5 mg.

Weigh a dry, stoppered weighing bottle 6 × 3 cm. diameter. Transfer to this bottle by means of a pipette an aliquot containing about 50 mg., and not more than 60 mg. P_2O_5. For example, if the material contains about 20 per cent P_2O_5, about 10 ml. is desirable. The volume of this aliquot, however, need not be measured accurately. Stopper the weighing bottle and weigh again to the nearest 0·5 mg.

---

* Some samples of phosphatic minerals are resistant to attack by hydrochloric acid, but perchloric acid dissolved all the phosphate from all the samples examined. This procedure also eliminates fluorine, and therefore precautions to prevent later attack on beakers (addition of boric acid) are unnecessary.

The weight of the sample actually taken for the analysis is then

$$\text{Total weight of sample} \times \frac{\text{Weight of solution in weighing bottle}}{\text{Weight of solution in flask}}.$$

(*iii*) Wash the stopper of the weighing bottle, collect the washings in a 500-ml. conical flask, and transfer the aliquot quantitatively to the flask. Wash the weighing bottle with about 90 ml. water, and add 20 ml. concentrated hydrochloric acid followed by 30 ml. sodium molybdate solution. Heat to boiling, and add a few drops quinoline solution from a burette with a coarse jet, whilst swirling the flask. Boil the solution again, and add 1–2 ml. quinoline solution dropwise to the gently boiling solution whilst swirling the flask. Boil yet again and add the reagent a few ml. at a time whilst swirling, until 60 ml. in all have been added. In this way a coarsely crystalline precipitate with good filtering properties is obtained. Place the solution in a bath of boiling water or on the edge of a hot-plate for 15 minutes and then cool to room temperature.

(*iv*) Prepare a paper-pulp filter in a funnel fitted with a porcelain cone, and clamp well down. Decant the clear solution through the filter, and wash the precipitate twice by decantation with about 20 ml. hydrochloric acid (1 : 9).* Transfer the precipitate to the pad with cold water, washing the flask well, and wash the filter and precipitate with small amounts of about 25–30 ml. cold water, letting each wash run through before adding the next, until the washes are acid-free when tested with litmus. Usually six washes are sufficient. Transfer the pad and precipitate back to the original flask (now acid-free). Insert the funnel in the flask and wash it well with water to ensure that all traces of the precipitate are transferred; use about 50 ml. water. Shake the flask well to break up the pad and precipitate completely.

(*v*) Run in exactly 50·0 ml. 0·5 N sodium hydroxide solution from a calibrated burette or pipette, swirling the flask during the addition. Shake until the precipitate is completely dissolved. Add a few drops indicator solution, and titrate with 0·5 N hydrochloric acid. The end-point is very sharp, the solution becoming pale green and changing suddenly to pale yellow.

Run a blank on all reagents excluding the aliquot of sample solution, but use 0·1 N solutions and calculate the result in terms of 0·5 N

---

* The preliminary washes with acid remove most of the excess of quinoline and molybdate, and prevent errors through the precipitation of quinoline molybdate or molybdic acid in the pad or precipitate.

sodium hydroxide solution. Subtract this blank from the volume neutralised by the original precipitate.*

$$1 \text{ ml. } 0\cdot5 \text{ N sodium hydroxide} \equiv 1\cdot366 \text{ mg. } P_2O_5.$$

$$\frac{\text{ml. } 0\cdot5 \text{ N sodium hydroxide} \times 0\cdot1366}{\text{sample weight analysed}} = \text{percentage } P_2O_5.$$

(*b*) *Determination of total P_2O_5 in mixed fertilisers containing ammonium salts.* (*i*) Weigh 2·500-g. sample into a 150-ml. beaker, add 1 g. boric acid and 20 ml. water, and warm until all the boric acid has dissolved. Add 10 ml. concentrated hydrochloric acid,† evaporate to dryness and bake gently for 15 minutes. Allow to cool, moisten the residue with 1–2 ml. concentrated hydrochloric acid, and add 15 ml. hot water. Warm until all the salts are dissolved, and filter as described in (*a,i*).

(*ii*) Proceed as described in (*a,ii*) above.

(*iii*) Transfer the aliquot to a 500-ml. flask as under (*a,iii*), but use only 70 ml. water. Add 2 g. solid sodium hydroxide (about 10 pellets) and swirl until the pellets have dissolved. This amount of sodium hydroxide will normally be sufficient; enough should be present to liberate all the ammonia from the salts and leave an excess. Add 20–40 ml. bromine water according to the amount of ammonia present (10 ml. bromine water destroys about 20 mg. ammonia). Mix the solution, allow it to stand for 5 minutes, acidify with concentrated hydrochloric acid added dropwise (normally 6 ml. acid is required) and boil gently to remove excess of bromine. Adjust the volume to 100 ml., add 20 ml. concentrated hydrochloric acid, and complete the determination as under (*a,iii*) above.

(*c*) *Determination of "water-soluble P_2O_5" in the absence of ammonium salts.* Weigh 10-g. sample into a 500-ml. volumetric flask (class A calibration) and add 40 ml. water at 20° C. Shake on a

---

* The blank determination is important. It is mostly due to silica and must be determined carefully. It should not be more than about 0·4–0·5 ml. About 0·2 ml. arises from the sodium molybdate, and the remainder from the sodium hydroxide and the glass apparatus. Soft soda or potash glass must not be used. Flasks that become scratched or etched must be discarded. Occasionally a flask that has been and is apparently satisfactory may begin to yield appreciable quantities of silica.

† This treatment is satisfactory for rendering silica insoluble. Perchloric acid is not used since the low solubility of ammonium perchlorate in water, and its rather unstable nature, may give rise to trouble. The addition of boric acid is essential if fluorides are present, and should precede acidification with hydrochloric acid.

machine for exactly 30 minutes.* Dilute to the mark with water, mix well, and filter through a dry Whatman No. 31 filter paper. Reject the first 20–30 ml., and then collect the filtrate. Adjust the temperature of the filtrate to 20° C. With a calibrated pipette (25- or 50-ml. according to the amount of soluble P_2O_5 expected) transfer a suitably sized aliquot to a 500-ml. conical flask. Dilute to about 100 ml. with water.

(*ii*) Add 20 ml. concentrated hydrochloric acid, and from this point proceed as under (*a,iii*) above.

(*d*) *Determination of "water-soluble P_2O_5" with ammonium salts present.* (*i*) Proceed as in (*c,i*) above.

(*ii*) Proceed to the destruction of ammonium salts as in (*b,iii*) above, and complete the determination as in (*a,iii*) above.

[1] H. N. Wilson, *Analyst*, 1951, **76**, 65.

19. GRAVIMETRIC DETERMINATION OF POTASSIUM AND SODIUM

To avoid the tedious separations normally required by the Lawrence Smith method for the determination of potassium and sodium in refractory materials, Haslam and Beeley[1] have described a method in which the leachings from the ignited mixture are made up to a known amount, and the potassium and sodium are determined directly on suitable aliquots. The determination may be completed on the following day, whereas the original method requires three days, even though one of the elements is determined by difference.

Potassium is precipitated as potassium sodium cobaltinitrite and weighed as such if the precipitate is small (say 12 mg.). Otherwise, because of the variable composition of the precipitate, this is

* "Water-soluble P_2O_5" has only one meaning in fertiliser analysis—as defined in the Fertiliser and Feeding Stuffs Act regulations. The regulations specify 20 g. sample and final dilution to 1 litre. Use of 10 g., as described, is satisfactory, and 500-ml. flasks are easier to accommodate on shaking machines. The time of shaking must be adhered to, and filtration must take place immediately because of possible slow reactions between the solid phase and the solution. Wilson did not find that silica in appreciable amounts passed into the solution during this treatment, but if a fertiliser of alkaline reaction were examined, silica might be dissolved.

redissolved, and the potassium is reprecipitated as the perchlorate and weighed in that form. Sodium is determined by precipitating as sodium zinc uranyl acetate.

Haslam and Beeley weigh the solution containing the leachings, and take known weights of this solution; but there seems to be no reason why the solution should not be made up to, say, 100 ml., and suitable volumes taken.

When potassium is present to the extent of 50 mg. per ml. the double potassium uranyl acetate is precipitated by the zinc uranyl acetate reagent. Kolthoff and Barber,[2] who developed the method for the determination of sodium, recommended that in such circumstances potassium should first be removed as the perchlorate. Also, since potassium sulphate is insoluble in the zinc uranyl acetate reagent, sulphates should be removed by addition of barium chloride. Haslam and Beeley applied this method to the determination of small amounts of sodium in potassium salts. They noted that barium perchlorate is insoluble in the zinc uranyl acetate reagent, and recommend that the excess barium ions remaining after sulphate removal be precipitated by ammonium carbonate before the addition of ammonium perchlorate to remove potassium.

This method may be applied to refractories containing a high proportion of potassium to sodium.

Solutions required

Sodium cobaltinitrite. Dissolve 113 g. cobalt acetate in a mixture of 300 ml. 35 per cent (v/v) acetic acid and 300 ml. water. Dissolve 220 g. sodium nitrite in 400 ml. water. Mix the two solutions and remove evolved gases by evacuation for 24 hours. Store in a cool dark place.

Zinc uranyl acetate. Dissolve 427 g. zinc acetate in a mixture of 46 g. 30 per cent (v/v) acetic acid and 527 g. water by heating on a water-bath. Similarly dissolve 154 g. uranyl acetate in a mixture of 92 g. 30 per cent (v/v) acetic acid and 748 g. water. Mix the hot solutions. Add 2 ml. 0·1 N sodium chloride solution, set aside for 24 hours, and filter.

Procedures

Opening up of refractory. Grind 1 g. finely powdered material in an agate mortar with 1 g. pure ammonium chloride and 6 g. pure calcium carbonate. Spread 0·5 g. calcium carbonate on the bottom of a deep

18

platinum crucible and place the prepared sample on top of this. Cover the mass with a further 0·5 g. calcium carbonate. Insert the covered crucible in a hole in an asbestos board, and heat with a small flame protected from draughts so that only ammonia vapour escapes. When the odour of ammonia is no longer perceptible (about 15 minutes) maintain the crucible at 800° C. for a further 40–60 minutes. Transfer the residue to a small porcelain basin and rinse the crucible thoroughly with hot water into the basin. Digest on the water-bath until the cake is completely disintegrated. If necessary, crush the cake with an agate pestle. Filter through a sintered-glass funnel into a 100-ml. flask, and wash the residue five times with 5-ml. portions of water. Make the filtrate just acid with acetic acid, and weigh.

Gravimetric determination of potassium. (a) As cobaltinitrite. Evaporate a known weight of the aqueous extract to about 25 ml. on the water-bath and cool to room temperature. Add 35 ml. cobalti-nitrite reagent and set aside for 4–5 hours or preferably overnight. If the precipitate is sufficiently small, filter on a Gooch crucible, wash with 10 per cent (v/v) acetic acid until the filtrate is colourless and then with 95 per cent (v/v) ethanol. Dry at 110° C. for 2 hours and weigh.

$$\text{mg. ppt.} \times 0{\cdot}2137 = \text{mg. } K_2O.$$
$$\text{mg. ppt.} \times 0{\cdot}1774 = \text{mg. } K.$$

(b) As perchlorate. If the amount of cobaltinitrite precipitate is large, filter on a Gooch crucible fitted with a small paper disc covered with paper pulp, and then wash with diluted cobaltinitrite reagent. Transfer the precipitate and filter paper to a small beaker and digest on the water-bath with dilute hydrochloric acid. Filter into a beaker, wash with water, and evaporate the solution to dryness to remove excess hydrochloric acid. Add a few ml. water to the dry residue, and 2–3 ml. 20 per cent perchloric acid solution. Evaporate on a sand-bath until fumes appear. After a few minutes remove the beaker and allow it to cool. Add further amounts of water and per-chloric acid and repeat the process. When the solution is quite cool add 25 ml. 98 per cent (v/v) ethanol, set aside for a few hours, filter on a Gooch crucible, wash with a saturated solution of potassium perchlorate in ethanol, dry at 100° C. and weigh as $KClO_4$.

$$\text{mg. ppt.} \times 0{\cdot}3399 = \text{mg. } K_2O.$$
$$\text{mg. ppt.} \times 0{\cdot}2822 = \text{mg. } K.$$

Gravimetric determination of sodium. Evaporate a known weight of the original solution (usually equivalent to 0·5 g. original sample) to dryness on the water-bath. Dissolve the residue in 2–3 ml. water, add 1 drop dilute acetic acid and add a volume of zinc uranyl acetate reagent at least ten times that of the water used to dissolve the residue. Set aside for 1 hour, filter on a Gooch crucible, wash five times with 2-ml. portions of the reagent solution and then five times with 2-ml. portions of 95 per cent (v/v) ethanol saturated with sodium zinc uranyl acetate. Finally wash with ether. Draw air through the crucible for 5 minutes, place the crucible in the balance case, and weigh after 1 hour.

$$\text{mg. ppt.} \times 0{\cdot}0201 \equiv \text{mg. Na}_2\text{O.}$$
$$\text{mg. ppt.} \times 0{\cdot}0149 \equiv \text{mg. Na.}$$

Determination of sodium in potassium chloride. Dissolve 1 g. potassium chloride in 4 ml. water and to the hot solution add 2 g. recrystallised ammonium perchlorate dissolved in 4 ml. water, followed by 25 ml. 95 per cent (v/v) ethanol. Cool the solution to room temperature. Filter on a X3 sintered glass funnel, and wash five times with 2-ml. portions of 95 per cent ethanol. Evaporate the filtrate to dryness, dissolve the residue in 3 ml. water at as low a temperature as possible, add 35 ml. zinc uranyl acetate reagent, and proceed as already described.

Determination of sodium in potassium sulphate. Dissolve 1 g. potassium sulphate in 50 ml. water, add 2 drops concentrated hydrochloric acid, and heat to boiling. Add 13 ml. barium chloride solution (containing 122·2 g. $\text{BaCl}_2 . 2\text{H}_2\text{O}$ per litre). When the sulphate is precipitated add 10 ml. ammonium carbonate solution (containing 2 g. ammonium carbonate, 1 ml. 0·88 ammonia and 9 ml. water), filter the mixed precipitates, and evaporate the filtrate to dryness. Dissolve the residue in 4 ml. water, remove the potassium as perchlorate and determine the sodium as already described.

[1] J. Haslam and J. Beeley, *Analyst*, 1940, **65**, 185.
[2] I. M. Kolthoff and H. H. Barber, *J. Amer. Chem. Soc.*, 1928, **50**, 1625.

20. DETERMINATION OF POTASSIUM AS POTASSIUM TETRAPHENYLBORON

The determination of potassium has always been one of the least satisfactory methods of quantitative analysis. Although many reagents have been proposed, all hitherto have had some disadvantage such as marked solubility of the precipitate, lack of selectivity, or doubtful composition of the precipitate. In general, the perchlorate method can be regarded as the most reliable,[1] but the procedure is rather lengthy.

Recently it has been shown that potassium can be precipitated quantitatively as the tetraphenylboron compound $K[B(C_6H_5)_4]$, which provides a very useful method for its determination. Under suitable conditions only ammonium, caesium and rubidium (which can also be determined) and a few other ions (e.g., Hg_2^{++} and Tl^+, which can readily be removed) interfere. From the amount of published work on this method which has appeared in a very short space of time, it would seem to be the most promising yet advanced. No doubt considerably more work requires to be done to devise a simple and accurate titrimetric finish, but it is felt that this text would benefit from inclusion of a general review of the present position.

Wittig and his co-workers[2] first discovered that the potassium salt was virtually insoluble, and gave details for a gravimetric determination. Raff and Brotz[3] also described a gravimetric method, and stated that the precipitate has a solubility only 0·1 per cent that of silver chloride. It contains 10·91 per cent potassium, and can be dried at 120°C. Conductimetric titration was also suggested, although the authors do not indicate if this suggestion was based on experiment. When alkaline earths are present they are precipitated by sodium carbonate before adding the reagent, and are subsequently redissolved in acetic acid. This treatment prevents coprecipitation.

Berkhout[4] modified the Raff and Brotz method, and was able to determine up to 20 mg. potassium in presence of 100 mg. dibasic calcium phosphate (by adding EDTA) or the same amount of ammonium chloride (by adding formaldehyde).

Kohler[5] has since questioned the solubility figure of Raff and Brotz. There has also been some disagreement on whether it is best to precipitate in the cold or at elevated temperatures, and on the most suitable medium. Neutral, acetic acid and dilute mineral acid media have been recommended. Raff and Brotz precipitate up to 100 mg. potassium in acetic acid medium in the cold. Flaschka,[6] working on the micro scale (39–395 μg. potassium) found that the precipitate crept badly in cold conditions, and heated to 70° C. to obtain a better product. He also found it necessary to wash with a solution saturated with the precipitate. As Kohler has pointed out, this should not be necessary if the precipitate is as insoluble as claimed by Raff and Brotz. Spier[7] also worked on the micro scale, but dissolved the precipitate in acetone and titrated with a perchlorate-cerate solution.

Various other titrimetric finishes have been investigated. Rüdorff and Zannier[8] tried bromometric and iodometric methods without success, but finally recommended an argentometric finish. The potassium salt is soluble in acetone, but the silver salt is not. To overcome slight solubility, small volumes must be used, and the final volume of titrant must not exceed that of the acetone. The colour change with eosin is poor, but a sharper end-point is obtained by adding a known amount of potassium bromide solution before titrating. The same authors state that lithium and sodium do not interfere, nor do magnesium or the alkaline earth metals in acetic acid medium. Aluminium and chromium do not interfere, but iron-III gives a dark precipitate. This is avoided by adding sodium fluoride to form the complex ion FeF_6^{---}. Normally potassium tetraphenylboron settles slowly and is difficult to filter, but by addition of a few drops of aluminium chloride solution rapid filtration and good settling are achieved, results which are also encouraged by precipitating at 70° C.

The most extensive examination to date is due to Kohler.[5] He reached the following conclusions:

(i) It is preferable to precipitate at room temperature in mineral acid solution because there are interferences in acetic acid solution and at elevated temperatures the reagent decomposes more readily. At room temperature in mineral acid the precipitate is coarse and readily filterable. Sulphuric, hydrochloric or nitric acid may be used. In 0·1 N mineral acid precipitation is effectively complete, in 0·5 N acid it is 92·3 per cent, and in 1·35 N acid it is 76 per cent.

(ii) Pure reagent is necessary.

(iii) The precipitate is more soluble than found by Raff and Brotz, and a wash solution saturated with precipitate is necessary.

(iv) Ammonium salts need not be fumed off, but can be precipitated with potassium and by suitable treatment both ions can be determined together.

(v) The precipitate can be filtered after a few minutes under the conditions recommended. Filtration should not be delayed beyond 90 minutes (preferably 60 minutes) owing to decomposition of excess reagent.

(vi) Under the recommended conditions, aluminium, calcium, cobalt, copper, iron-II, iron-III, magnesium, manganese, nickel, sulphate and phosphate do not interfere. Only mercury-I, which is rarely encountered, ammonium, caesium and rubidium precipitate.

Flaschka and his co-workers[9] tried to develop a titrimetric method based on the reaction (recorded by Wittig and Raff)

$$K[B(C_6H_5)_4]+4HgCl_2+3H_2O \longrightarrow$$
$$4C_6H_5HgCl+3HCl+KCl+H_3BO_3.$$

It was hoped to titrate the hydrochloric acid formed or the hydrochloric acid plus boric acid, but the reaction did not prove suitable. They recommended ignition of the precipitate to potassium *meta*borate in a platinum crucible, followed by acidimetric titration. If ammonium is present this is precipitated also. On heating, the ammonium salt volatilises leaving only the potassium *meta*borate. The ignition method is particularly suitable for biological fluids, where ashing is essential since it is not known whether the reagent carries down organic matter or not.

Later the same workers[10] found a means of making the reaction with mercuric chloride workable, and developed a further titrimetric method based on this reaction. The main difficulties in utilising the reaction lay in the interference from mercury phenyl chloride. This was overcome by adding excess potassium iodide to form the more strongly bound iodide complex. The reaction proceeds rapidly in alkaline solution, hence an excess of standard alkali is added, followed by back titration with hydrochloric acid.

These authors also found the device of Rüdorff and Zannier, *i.e.*, granulation of the precipitate with a trace of aluminium chloride, useful. Although aluminium hydroxide formed at the pH of the precipitation is converted to aluminate ion when excess standard alkali is added, aluminium hydroxide is reprecipitated at the end-point on back titration with hydrochloric acid using methyl red as indicator, and so takes no part in the overall reaction.

The washing technique requires care. The ion $[B(C_6H_5)_4]^-$ is the entity being titrated, and this gives 3HCl per mole. When a saturated wash solution is used on a precipitate which is damp with excess reagent ($Na[B(C_6H_5)_4]$) the common ion precipitates the potassium compound from the wash solution. It is therefore necessary to wash first with a few drops of water, then with wash solution, and finally with a few drops of water.

The method has been applied over the range 0·03–3·0 mg. potassium, and appears to be the best titrimetric finish yet advanced.

The general method has already found some application, having been used to determine potassium in milk,[11] artificial manures,[12] biological fluids[7, 12] and medicinals.[8, 13] A selection of the various procedures is given below.

Solutions required

Wash solution of Flaschka. Dissolve dry potassium tetraphenylboron in acetone to give 15·0 mg. per ml. Add 4 ml. solution to 100 ml. water, producing a nearly saturated solution.

Wash solution of Kohler. Precipitate about 0·1 g. potassium from a 0·1 N hydrochloric acid solution with pure reagent, filter under reduced pressure after a few minutes, and store the dry salt over calcium chloride in a desiccator, or prepare fresh every few days. Shake 20–30 mg. of this material with 250 ml. distilled water for 30 minutes, add 0·5–1·0 g. aluminium hydroxide, and stir for a few minutes. Filter on a Whatman No. 1 filter paper, rejecting or refiltering the first 20 ml. filtrate.

Reagent solution. Dissolve 1·5 g. commercial sodium tetraphenylboron in 250 ml. distilled water, and add 0·5–1·0 g. pure, alkali-free aluminium hydroxide. Stir for 5 minutes, filter through a Whatman No. 1 filter paper, and reject or refilter the first 20 ml. filtrate. This provides a 0·6 per cent solution of the reagent. Prepare fresh every few days.

Procedures

Method of Raff and Brotz. The solution should contain up to 100 mg. potassium in 100–200 ml. Make acid with acetic acid, and add whilst stirring a 50 per cent excess of the sodium (or lithium) salt of tetraphenylboron. After 5 minutes filter on a filter crucible, wash with very dilute acetic acid, dry at 120° C. and weigh.

If alkaline earth metals are present, precipitate these with sodium carbonate, add the reagent, and redissolve the alkaline earth carbonates with dilute acetic acid.

$$\text{mg. ppt.} \times 0.1091 \equiv \text{mg. K.}$$

Method of Rüdorff and Zannier. Add a few drops of 0.2 N aluminium chloride solution to the potassium solution, and adjust to pH 5.0–6.0 with acetic acid or, if already acid, with sodium acetate. (In mineral acid solution the sodium salt separates out.) Heat to 70°C., and slowly add a 50 per cent excess of the 0.1 N reagent in water. If iron is present add solid sodium fluoride till decolorised. Cool, and filter on a Whatman No. 1 filter paper. Wash with dilute acetic acid until the washings no longer precipitate with silver nitrate. Transfer the paper and precipitate to the original beaker, dissolve the precipitate in 10 ml. acetone per 25 mg. potassium. Add 5 ml. 2 N acetic acid, 1 ml. 0.1 N potassium bromide solution, 2 drops 1 per cent eosin indicator, and titrate with 0.05 N silver nitrate solution (0.1 N if the potassium content is more than 25 mg.) till the indicator changes colour. Deduct the correct amount for the potassium bromide.

$$1 \text{ ml. } 0.1 \text{ N silver nitrate} \equiv 3.9096 \text{ mg. K.}$$

Rubidium can be determined by following exactly the same procedure.

Method of Kohler. For amounts of about 10 mg. add sufficient acid to make the solution 0.1 N when diluted to 50 ml. Add 20 ml. 0.6 per cent reagent solution while swirling the beaker (in general, 10 ml. 0.6 per cent reagent per 5 mg. K_2O), and make the volume up to 50 ml. Filter on a fine porcelain or sintered-glass filter crucible, wash with water saturated with precipitate, dry at 105°C. for 30 minutes, cool in a desiccator, and weigh.

For amounts of the order of 20 mg. use 40 ml. reagent and dilute to 100 ml.; retain a final acidity of 0.1 N.

Good results are obtained down to 2 mg. potassium or ammonium. If the potassium content is less than 2 mg. a known amount of potassium should be added (3–5 mg.) and this amount then deducted from the final result.

When potassium and ammonium are both present both are precipitated. After weighing the precipitate dissolve it from the filter with 10–20 ml. acetone, filter into a 250 ml. beaker, and treat with

20 ml. 2·5 per cent sodium hydroxide solution free from potassium. Evaporate the solution on the water bath to volatilise the acetone and the ammonium salt. Cool, add water, and filter off the potassium salt. Wash this, dry at 105° C., cool and weigh. The ammonium content is given by difference.

Method of Flaschka, Holasek and Amin. Adjust the test solution, contained in a roomy platinum crucible, to pH 4·0–6·0 with 3 per cent acetic acid. Add 0·5–2·0 ml. 0·3 per cent ammonium chloride solution. This helps to give a coarser precipitate, and is useful when only small amounts of potassium are present. Heat to boiling, remove from the source of heat, add 1·5–2·0 ml. excess 3 per cent reagent solution, and allow to cool to room temperature. Filter with a porcelain filter stick, and wash with a small amount of water. (Two or three portions of 0·5–1·0 ml.) Dry the precipitate, add 1·0–2·0 ml. acetone, and help solution by rotating and scraping with the filter stick. Rinse the filter stick with a small amount of acetone introduced through the tube. Evaporate off the acetone under an infra-red lamp, char over a small flame, and then heat to redness and cool.

Add 1 ml. 0·01–0·001 N hydrochloric acid from a pipette, heat to boiling, and back titrate with standard sodium hydroxide solution of similar strength, using methyl red-methylene blue indicator.

A porcelain or quartz crucible may be used instead of a platinum one, but the precipitate should be ignited at as low a temperature as possible.

$$(\text{ml. N NaOH} - \text{ml. N HCl}) \times 39 \cdot 096 \equiv \text{mg. K.}$$

Alternative method of Flaschka and Co-workers. Treat the neutral solution with 1 drop 1 per cent acetic acid. Heat to 70°–80° C. Stir, and precipitate by adding 3 per cent aqueous solution of sodium tetraphenylboron in at least one and a half to two-fold excess. The excess does not interfere. Add 1–2 drops 0·02 N aluminium chloride solution, and allow to cool while stirring. Filter through a porcelain filter stick and suck dry. Wash with 5–6 drops water, then with 1–2 ml. wash solution and finally with 2–3 drops water. Dissolve the precipitate in 2–4 ml. acetone, washing the precipitate from the sides of the vessel and the plate of the filter stick. Add 1–2 ml. saturated mercuric chloride solution (according to the amount of precipitate), a known excess of 0·01 N sodium hydroxide solution and 0·02 per cent ethanolic methyl red indicator solution. Heat to boiling. If the indicator turns red, add more 0·01 N sodium hydroxide solution. After a short period of boiling add dropwise 20 per cent potassium iodide solution until all the mercury is precipitated. Titrate with 0·01 N hydrochloric acid solution to a pronounced red colour, boil to expel carbon dioxide, and bring back to the neutral-point with 0·01 N sodium hydroxide.

$$(ml.\ 0·01\ N\ NaOH - ml.\ 0·01\ N\ HCl) \times 0·13032 \equiv mg.\ K.$$

Method of Berkhout. For amounts of potassium up to 20 mg. in 100 ml. solution, in the presence of up to 100 mg. ammonium chloride of calcium phosphate, add 1 drop phenolphthalein indicator solution, 10 ml. 4 per cent EDTA solution, and a 30 per cent sodium hydroxide solution dropwise until the solution is red. Heat to boiling, add 5 ml. 25 per cent formaldehyde solution, and add dropwise 10 ml. 3 per cent sodium tetraphenylboron solution. The solution should be alkaline after the addition of formaldehyde. Cool, filter under suction and wash with cold water. Dry for 30 minutes at 120°C., and weigh.

Method of Spier. Ash the residue from a biological fluid, and carry out the precipitation as in the method of Raff and Brotz. Dissolve the dried precipitate in acetone, heat for 15 minutes at 92°–97°C., and titrate with 0·025 N ammonium hexanitratocerate in 2 N perchloric acid.

$$1\ ml.\ 0·025\ N\ cerate \equiv 0·00135\ mg.\ K.$$

[1] H. H. Willard and H. Diehl, *Advanced Quantitative Analysis*, D. Van Nostrand Co., Inc., New York: 1944, p. 253.
[2] G. Wittig, G. Keicher, A. Rückert and P. Raff, *Liebigs Ann.*, 1949, **563**, 114, 118, 126: G. Wittig, *Angew. Chem.*, 1950, **62**, 231: G. Wittig and P. Raff, *Liebigs Ann.*, 1950, **573**, 195.
[3] P. Raff and W. Brotz, *Z. anal. Chem.*, 1951, **133**, 241.
[4] H. W. Berkhout, *Chem. Weekblad*, 1952, **48**, 909.
[5] M. Kohler, *Z. anal. Chem.*, 1953, **138**, 9.
[6] H. Flaschka, *ibid.*, 1952, **136**, 99.
[7] H. W. Spier, *Biochem. Z.*, 1952, **322**, 467.
[8] W. Rüdorff and H. Zannier, *Z. anal. Chem.*, 1952, **137**, 1.
[9] H. Flaschka, A. Holasek and A. M. Amin, *ibid.*, 1953, **138**, 161.
[10] *Idem, ibid.*, 241.
[11] I. R. Schober and A. Fricker, *Z. Lebensm. Unt.*, 1952, **95**, *C.A.* 107; 1952, **46**, 11479.
[12] J. Schwaibold and M. Kohler, *Landwirtschaffliches Jahrbuch für Bayern*, 1953, **30**, 1–2, 55.
[13] O. E. Schulz and O. Mayer, *Deutsche Apotheker Z.*, 1952, **21**, 358.

21. TITRIMETRIC DETERMINATION OF SILICON

(a) AFTER PRECIPITATION AS POTASSIUM FLUOSILICATE

Normally silicon in iron and steel is determined gravimetrically or, in certain cases, colorimetrically. Kordon[1] has recently developed a titrimetric method based on precipitation of the silicon as potassium fluosilicate, followed by alkalimetric titration of the precipitate according to the equation:

$$K_2SiF_6 + 4KOH \longrightarrow 6KF + Si(OH)_4.$$

The method is not new; for example, a method based on a similar principle was described by Travers[2] in 1926. But Kardon, for the first time, has studied and eliminated the main sources of error in the method.

In order to reduce the solubility of the precipitate, the precipitation is effected in the presence of excess potassium chloride and hydrofluoric acid. The presence of the latter necessitates the use of synthetic resin apparatus.

It was established electrometrically that when potassium fluosilicate is titrated with alkali the equivalence-point occurs at a pH of 7·1; but when bromothymol blue, the most suitable indicator for this range, was used, the end-point was masked by a grey colour imparted to the solution by graphite. The change with phenolphthalein can be detected readily, but 0·35 ml. of the standard alkali (1 ml. ≡ 1·0 per cent Si on a 1 g. sample) is consumed above that required by theory. This may be overcome quite readily by deducting 1·0 per cent from the silicon value obtained. When the silicon content of the sample is below 1·0 per cent, as is the case with unalloyed and most alloyed steels, the correction can be omitted.

The titration must be carried out at high temperature, otherwise the reaction between the potassium fluosilicate and the potassium hydroxide is incomplete.

With alloyed steels, it is found that aluminium, tantalum, titanium and zirconium interfere owing to the formation of potassium fluo-salts which are almost insoluble under the experimental conditions employed. Their effect cannot be eliminated by increasing the concentration of acid, by washing with acid, or by the addition of organic acids. However, as long as these elements do not exceed 0·5 per cent of the steel, satisfactory results are obtained by the procedure as described.

Sulphuric acid must be avoided, as low results are obtained in its presence. Phosphoric acid is without effect.

Solutions required

Hydrochloric-phosphoric acid. To 1000 ml. hydrochloric acid (sp. gr. 1·1) add 100 g. potassium dihydrogen phosphate dihydrate.

Standard alkali. Either 0·1 N NaOH or alternatively a solution containing 5·7021 g. NaOH per litre.

Procedures

Titrimetric determination of silicon in pig iron or in carbon steels.

Transfer 1·0 g. to a 250-ml. beaker and dissolve in 20 ml. hot acid (pig iron in hydrochloric acid, sp. gr. 1·19; steel in nitric acid, sp. gr. 1·2) on the hot-plate. When nitric acid is used add 10 ml. hydrochloric acid when solution is complete, and boil for 1 minute to dissolve insoluble compounds formed during the nitric acid attack on the steel. Otherwise these will be filtered off and will consume some alkali. Cool rapidly to 30° C. and transfer to a 250-ml. resin beaker. Stir continuously with a resin stirrer while adding potassium chloride until the solution is saturated (5–7 g. solid potassium chloride). Add 5 ml. hydrofluoric acid. Stir for a further 3 minutes, and filter through a paper-pulp pad supported on a resin or platinum filter disc contained in a resin funnel. Wash with the minimum amount of 20 per cent potassium chloride solution until the washings are acid-free.

Transfer the pad and precipitate to a 500-ml. conical flask, add 100 ml. carbon dioxide-free boiling water, shake thoroughly, and titrate while still hot with standard alkali solution, using phenolphthalein indicator, until a pink colour appears. Boil for a few seconds, and if the colour fades continue the titration until a permanent colour is obtained. Deduct the value obtained on conducting a blank test on the reagents.

1 ml. 0·1 N sodium hydroxide   ≡0·7015 mg. Si.
1 ml. sodium hydroxide (5·7021 g. per l.)   ≡0·1 per cent Si
                                                                      on a 1 g. sample.

Titrimetric determination of silicon in tungsten-free alloy steels.
Dissolve 1·0 g. chips in 20 ml. nitric acid (sp. gr. 1·2) on the hot-plate,

and add 10 ml. hydrochloric acid (sp. gr. 1·19) when solution is complete. For higher alloy steels with carbide-bearing elements, *e.g.*, with more than about 4 per cent chromium and/or vanadium, dissolve in 20 ml. hydrochloric acid (sp. gr. 1·19) and oxidise by the gradual addition of 10 ml. nitric acid (sp. gr. 1·2). Then follow the procedure already described for pig iron.

Titrimetric determination of silicon in tungsten-bearing alloy steels. Dissolve 1·0 g. chips in 40 ml. hydrochloric-phosphoric acid on the hot-plate. Oxidise by adding 10 ml. 5 per cent potassium chlorate solution, and allow to stand for 2 minutes at a fairly high temperature to bring the carbide into solution. Add 30 ml. hot water, boil for 2 minutes, and continue as before.

[1] F. Kordon, *Archiv. Eisenhüttenw.*, 1945, **18**, 139.
[2] A. Travers, *Compt. rend.*, 1927, **185**, 893.

(b) After Precipitation as Quinoline Silicomolybdate

Silicon, in the form of *ortho*silicic acid, may be precipitated as the quinoline salt of silicomolybdic acid, $(C_9H_7N)_4H_4(SiO_4 . 12MoO_3)$. This salt may then be dissolved in standard sodium hydroxide solution, breaking down the complex to quinoline, silicic acid and molybdic acid. The free quinoline and silicic acid are too weak to react, and only the molybdic acid reacts with the sodium hydroxide according to the equation:

$$(C_9H_7N)_2H_4(SiO_4 . 12MoO_3) + 24NaOH \longrightarrow$$
$$4C_9H_7N + SiO_2 + 14H_2O + 12Na_2MoO_4.$$

The excess of sodium hydroxide may then be determined, and each equivalent of sodium hydroxide used corresponds to 1/24 the molecular weight of silica. As a consequence, the theoretical equivalent of silica by this method is 2·5025. Over a large number of experiments[1] the value found was 2·513.

The silica-bearing material is first carefully fused with solid sodium hydroxide. As this reaction may be violent, care must be taken to avoid loss of silica. The melt is extracted with water, and the extract is poured into strong hydrochloric acid. Practically all of the silica remains in solution as *ortho*silicic acid. The solution is then made alkaline, and the pH adjusted to 1·5 in order to take into solution any decomposed *ortho*silicic acid. A moderate excess of ammonium molybdate then converts the *ortho*silicic acid to silicomolybdic acid quantitatively. The solution, made even more acid to prevent precipitation of quinoline molybdate, is treated with a solution of quinoline hydrochloride, which precipitates the complex.

The corresponding pyridine salt is too soluble, and neither naphthylamine nor aniline form insoluble salts. The hexamethylenetetramine salt recommended by Duval[2] gives trouble on the macro scale owing either to decomposition of the complex or to hydrolysis of the hexamethylenetetramine, unless the pH is very rigidly controlled. The 8-hydroxyquinoline salt recommended by Brabson and his co-workers[3] gives excellent results for a gravimetric finish, but cannot be satisfactorily decomposed to permit a titrimetric procedure to be applied.

The method is suitable for amounts of silica up to 60 mg., representing about 12 per cent of a 0·5 g. sample. It has been applied successfully to a few common siliceous materials, but in general interferences have not been investigated. Phosphorus, however, should be absent.

The standard deviation was found to be 0·048, as compared with a value for a similar class of material, using the classical method, of 0·083. The method is therefore more accurate than the classical method. It is also more rapid, since the whole procedure, including the fusion (but not the necessary blank determination on the reagents) may be carried out in 90 minutes.

Solutions required

Thymol blue indicator solution. Dissolve 0·4 g. thymol blue in 200 ml. ethanol, add 8·6 ml. 0·1 N sodium hydroxide solution, and dilute with water to 1 litre.

Cresol red-thymol blue indicator solution. Grind 0·1 g. cresol red with 5·3 ml. 0·1 N sodium hydroxide until the indicator is dissolved. Dilute with water to 100 ml. Dissolve 0·1 g. thymol blue in 20 ml. ethanol, add 2·1 ml. 0·1 N sodium hydroxide solution, and dilute with water to 100 ml. Mix the two solutions.

Procedure

Titrimetric determination of silica. Grind the sample in an agate mortar, and if necessary burn off carbonaceous material in a muffle furnace for 15 minutes at 950°–1000°C. Weigh 7 g. AnalaR sodium hydroxide into a 5-cm. nickel crucible, and fuse to drive off water. Cool and add the weighed sample. Fuse gently and carefully for 3 minutes, raising to dull red heat at the end of the fusion. Allow to cool nearly to room temperature, and place in a covered 500-ml. tall-form beaker. Fill the crucible with water at 100°C. from a wash-bottle. The heat of solution should just keep the water boiling, but should not cause splashing. Allow the vigorous reaction to subside, wash down the cover-glass and beaker, and wash and remove the crucible. Give a final rinse to the inside of the beaker with several ml. warm hydrochloric acid (1 : 1), collecting these acid washings separately in a 500-ml. stoppered conical flask which is marked to indicate a volume of 170 ml. Add 20 ml. concentrated hydrochloric acid to this flask, and pour the solution and washings from the beaker into this, with constant agitation. Wash the inside of the beaker with hot water. Heat the flask on a hot-plate to redissolve any precipitate, but do not allow the solution to boil. Cool rapidly, and make up to the 170-ml. mark with water. Add 3 g. pellet sodium hydroxide, and when this has dissolved follow with 8 drops thymol blue indicator solution. Add concentrated hydrochloric acid drop by drop, with thorough mixing, till the colour just changes to red.

Add 8 ml. hydrochloric acid (1 : 9), 5 ml. acetic acid (1 : 2) and 30 ml. 10 per cent filtered aqueous ammonium molybdate solution which should not be more than a week old. Keep the solutions well agitated throughout, and shake vigorously for a further minute at the end. Immerse the flask in a boiling water-bath to raise the temperature of the solution to 80°–90°C. over 10–12 minutes. Remove the flask from the bath, add 40 ml. hydrochloric acid (1 : 1) and immediately run in from a fast burette 65 ml. of a solution prepared by dissolving 20 ml. quinoline (b.p. 230°–240°C.) in 800 ml. hot hydrochloric acid (1 : 32) with constant stirring, cooling to room temperature and adding paper pulp, filtering when the pulp has settled, and making up to 1 litre with water. Replace the flask in the water-bath and keep its contents at 80°–90°C. for a further 5 minutes, with occasional swirling. Cool rapidly to 15°C. and set aside until the precipitate has settled.

Decant the supernatant liquid through a paper-pulp pad, and wash the precipitate twice by decantation with 25–30 ml. cold water. The wash water should be kept below 15°C. Transfer the precipitate to the pad and wash thoroughly with six 30-ml. portions of cold water. Transfer the precipitate and pad to the original conical flask, and run in 30 ml. N sodium hydroxide solution (carbon dioxide-free). Wash any traces of precipitate from the funnel into the flask with the minimum amount of water. Stopper the flask and shake vigorously until all the yellow precipitate has completely dissolved. Add a few drops cresol red-thymol blue indicator solution, and back titrate with 0·5 N hydrochloric acid until the indicator just turns yellow.

Carry out a blank determination, for which the usual value is about 0·6 ml., owing largely to silica in the sodium hydroxide. Deduct the blank from the main titration.

$$1 \text{ ml. N sodium hydroxide} \equiv 2 \cdot 503 \text{ mg. } SiO_2.$$

[1] H. N. Wilson, *Analyst*, 1949, **74**, 243.
[2] C. Duval, *Anal. Chim. Acta*, 1947, **1**, 33.
[3] J. A. Brabson, H. C. Mattraw, G. E. Maxwell, A. Darrow and M. F. Needham, *Analyt. Chem.*, 1948, **20**, 504.

22. GRAVIMETRIC DETERMINATION OF SODIUM

The conventional double acetate reagents for the precipitation of sodium as $NaM(UO_2)_3(CH_3.COO)_9.6H_2O$ (p. 256) also precipitate lithium when it is present in amounts greater than 1–2 mg. (p. 243), necessitating its prior removal as the fluoride. Caley and Rogers[1] have developed a new double acetate reagent—copper uranyl acetate —which is insensitive to lithium.

Amounts of sodium between 1 and 50 mg. may be determined satisfactorily. Lithium causes slight positive errors, but these are not serious up to about 10 mg. lithium. The error is decreased slightly by increasing the volume of the test solution. The error increases with increase in the amount of sodium precipitated.

Interferences with the reagent are the same as those occurring with the well-established double acetate reagents, the only notable difference being ammonium, which causes slightly low results when present in considerable amounts. This ion, however, is readily eliminated.

An ethanolic reagent is used since it is more sensitive to sodium and more insensitive to lithium than the aqueous reagent. It is stable for as long as a year if kept out of strong light, and should be stored in a dark bottle.

From a manipulative standpoint an ethanolic reagent is less convenient than an aqueous reagent, because the precipitates are more difficult to transfer. The present reagent is only recommended, therefore, when lithium is present.

Solution required

Copper uranyl acetate solution. Dissolve 40 g. uranyl acetate dihydrate and 25 g. cupric acetate monohydrate in 450 ml. water and 100 ml. glacial acetic acid at a temperature of 50°–60°C. Cool to room temperature and add 500 ml. 95 per cent ethanol with constant stirring. Set aside for 2–3 days, stir, and filter through a dry filter.

Procedure

Gravimetric determination of sodium. Reduce the volume of the neutral solution to 5 ml. and add a volume of reagent corresponding to the probable sodium content. Not less than 100 ml. reagent should be used when the sodium content is 10 mg. or less. For larger amounts the number of ml. reagent solution should be ten times the number of mg. sodium present.

Immerse the vessel in a water-bath maintained at 20°C., and stir vigorously for 30–45 minutes. Filter on an X3 sintered-glass filter crucible, and when all the solution has run through, wash successively with 5-ml. portions of 95 per cent ethanol freshly saturated with sodium copper uranyl acetate.

Dry to constant weight at 60°C., and weigh.

$$\text{mg. ppt.} \times 0\cdot0145 \equiv \text{mg. Na.}$$

[1] E. R. Caley and L. B. Rogers, *Ind. Eng. Chem. Anal.*, 1943, **15**, 32.

23. TITRIMETRIC DETERMINATION OF THORIUM

Banks and Diehl[1] have described a method for the determination of thorium based on precipitation as thorium molybdate. The correct amount of precipitant to be added is determined by using diphenylcarbazide as indicator. After the precipitate is isolated the molybdenum is reduced and titrated oxidimetrically. The precipitation is effected in the presence of acetic acid, calcium not being precipitated under these conditions as long as not more than 0·4 g.

19

is present. Larger amounts of calcium cause high results. The non-precipitation of calcium is thought to result from the incomplete dissociation of calcium acetate.

Although uranyl molybdate precipitates in the presence of acetic acid, ammonium acetate prevents its precipitation for several hours, even if the solutions are boiled. A larger excess of precipitant is necessary, however, and diphenylcarbazide cannot be used to show when precipitation is complete. It is necessary, therefore, to carry out a preliminary determination to ascertain the correct amount of precipitant to use, or else to test the supernatant liquid.

It is not possible to determine thorium in the presence of the lanthanons by this method.

An electrometric titration procedure was also developed, in which the thorium solution is titrated with standard ammonium *para*-molybdate solution, using a calomel reference electrode and a molybdenum wire indicator electrode.

Solutions required

Ammonium paramolybdate solution. Dissolve 7·6 g. in water and make up to 1 litre.

Diphenylcarbazide indicator. Dissolve 0·5 g. in 200 ml. 95 per cent ethanol.

Procedures

Titrimetric determination of thorium. Transfer the sample, containing 0·15–0·2 g. thoria, to a 250-ml. beaker. Dissolve in a suitable mineral acid and evaporate nearly to dryness to remove excess acid. Dilute to about 150 ml. with water, and make about 7 per cent to acetic acid by adding 11 ml. glacial acetic acid. Add 15 ml. paper-pulp slurry and 1 ml. diphenylcarbazide indicator solution. Add ammonium molybdate solution from a burette until the indicator imparts a deep pink colour to the solution. Allow the precipitate to settle, and test the supernatant liquid for complete precipitation.

Heat the contents of the beaker to boiling, and filter hot through a Whatman No. 42 filter paper. Wash the precipitate five to six times with hot acetic acid (1 : 100). It is not necessary to remove the last traces of the precipitate from the beaker, and two to three rinses with wash solution are sufficient. Complete the determination as already described for molybdenum (p. 245).

1 ml. 0·1 N ceric sulphate ≡ 3·689 mg. Th.

Titrimetric determination of thorium in the presence of uranium. Transfer the sample, containing 0·15–0·2 g. thoria and not more than 0·5 g. uranium, to a 250-ml. beaker. Dissolve and evaporate as described above. Add 5 g. ammonium acetate and 11 ml. glacial acetic acid, dilute to 150 ml., and complete the determination as before.

Analysis of thorium-uranium alloys. Decompose thorium-uranium alloys with hydrochloric acid and bring them into solution by fuming with perchloric acid.

Determine the uranium on a separate sample by reducing in a Jones reductor, aerating, and titrating with 0·1 N ceric sulphate solution.

[1] C. V. Banks and H. Diehl, *Ind. Eng. Chem. Anal.*, 1947, **19**, 222.

24. TITRIMETRIC DETERMINATION OF ZINC

Zinc may be determined by addition of excess standard potassium ferrocyanide solution, and titration of the excess with standard ceric solution, using ferrous-*o*-phenanthroline as indicator.[1]

Cadmium, manganese and tin do not interfere seriously, and interference from iron may be reduced by addition of potassium ferricyanide. Each mole of iron-II present produces one mole ferrocyanide (which is precipitated by an equivalent amount of the zinc present in the solution) and one mole of iron-III (which remains in solution).

$$Fe^{++}+Fe(CN)_6^{---} \longrightarrow Fe^{+++}+Fe(CN)_6^{----}.$$

The iron-III subsequently reacts with the added standard ferrocyanide, but this time in the ratio of 4 : 3:

$$4Fe^{+++}+3Fe(CN)_6^{----} \longrightarrow Fe_4[Fe(CN)_6]_3.$$

As a consequence, each mole of iron-II originally present will leave 0·25 mole ferrocyanide in excess, and will cause the final result for zinc to be low by that amount.

Acid concentration is kept high to avoid interference from small amounts of cadmium and manganese. If tin-II is present its effect must be eliminated by addition of mercuric chloride, and a potentiometric finish is necessary.

Ammonium salts, nitrates or a high sulphate content interfere with the titration.

Solutions required

Standard (0·025 N) potassium ferrocyanide. Dissolve 0·2 g. sodium carbonate in 1 litre water and add 11·2 g. potassium ferrocyanide trihydrate. Allow to stand overnight, filter, and standardise against zinc chloride solution made by dissolving 1 g. zinc in 20 ml. hydrochloric acid (1 : 1) and diluting to 1 litre with water.

Ammonium sulphatocerate solution (0·025 N). Dissolve 22 g. dihydrate in approximately N sulphuric acid and filter. Standardise against the ferrocyanide solution.

Indicator solution. Dissolve 1·485 g. 1 : 10-phenanthroline in 100 ml. 0·025 M ferrous sulphate solution.

Procedure

Titrimetric determination of zinc in magnesium alloys. Weigh a 2-g. sample of the alloy into a 400-ml. beaker, and add 75 ml. distilled water and 25 ml. concentrated hydrochloric acid. If copper is present, add 3 g. pure lead metal. Cover the beaker and boil for 5 minutes. Decant the solution through a Whatman No. 1 filter paper into a 600-ml. beaker, and wash the residue with three 10-ml. portions of water. Cool to 45° C., and add 2 ml. 0·5 per cent aqueous potassium ferricyanide to deal with iron interference. Add a known excess of standard ferrocyanide solution, dropwise at first, and stirring mechanically to prevent formation of a colloidal precipitate which might pass through the filter. Gradually increase the rate of addition after the first few ml. Set aside for 5 minutes, and filter by suction through a Whatman No. 42 filter paper covered with a thin layer of asbestos into a 500-ml. suction flask. Wash with two 15-ml. portions of hydrochloric acid (1 : 10). Add 2 drops indicator solution, and titrate with the standard cerate solution. Deduct 0·2 ml. from the titration figure as an indicator correction.

1 ml. 0·1 N potassium ferrocyanide≡6·538 mg. Zn.

[1] L. G. Miller, A. J. Boyle and R. B. Neill, *Ind. Eng. Chem. Anal.*, 1944, **16**, 256.

SUBJECT INDEX

Abbreviations

E = extraction
GD = gravimetric determination
I = indicator
P = precipitation
R = reagent

So = solvent
St = standardisation
Sy = synthesis
TD = titrimetric determination

ACETONYL acetonates, **E**, 41
Acid-base indicators, 128
Acid, **St**, *tris*-(hydroxymethyl)-amino-
   methane, 105
Adsorption indicators, 157
Alizarin blue, **R**, 76
Alphazurin-methyl red, **I**, 137
Aluminium chloride, **R**, 230
   St, 230
Aluminium, **GD**
   N-benzoylphenylhydroxylamine,
      82
   8-hydroxyquinoline, 114
   P, benzoate, 24
   TD, potassium bromate, 115
Amaranth, **I**, 149, 150
5-Amino-2 : 3-dihydro-1 : 4-phthal-
   azinedione, **I**, 168
2-Amino diphenylamine-4-sulphonic
   acid, **I**, 144
Ammonium benzoate, **R**, 19
Ammonium hexanitratocerate, **R**, 211,
   266
   St, 212
   molybdate, **R**, 250, 274
   oxalate, **R**, 215
   phosphomolybdate, **I**, 190
   sulphatocerate, **R**, 276
   sulphatocerate-starch, **I**, 166
Ammonium, **TD**, hypochlorite, 182
Amyl acetate, **So**, 37
Amylose, **I**, 156
Antimony, **E**, 34
   GD, *trans*-Dichloro-*bis*-ethylene-
      diamine cobalt-III chloride,
      68
   TD, iodate, 177
   percuprate, 195
Arsenic, **TD**, iodate, 175
   percuprate, 195

BARIUM, **P**, nitrate, 6
   sulphate, 31
   TD, iodate, 47
   sodium thiosulphate, 47

Basic formates, **P**, 8
Benzene-2-hydroxybenzanthrone, **I**,
   133
Benzene-2-oxobenzanthrone, **I**, 133
Benzidine acetate, **I**, 146
Benzidine-copper complex, **I**, 161
Benzoates, **P**, 19
α-Benzoinoxime, **R**, 228
5 : 6-Benzoquinoline, **R**, 78
Benzo-1 : 2 : 3-triazole, **R**, 79
Benzoyl auramine G, **I**, 131
N-Benzoylphenylhydroxylamine, **R**, 81
   Sy, 83
Beryllium, **GD**, ethylenediaminetetra-
   acetic acid, 98
Bismuth, **GD**, basic formate, 10
   ethylenediaminetetra-acetic acid, 97
Bordeaux B, **I**, 148, 151
Brilliant Ponceau 5R, **I**, 150, 151
Bromate, **TD**, stannous chloride, 191
Bromide, **TD**, sodium thiosulphate,
   201
Bromocresol green-methyl red, **I**, 138,
   139
   methyl yellow, **I**, 138
   sodium alizarin sulphonate, **I**,
      105
Bromocresol purple, **I**, 158
p-Bromomandelic acid, **R**, 119
Bromophenol blue, **I**, 158
Bromothymol benzein-phenol red, **I**,
   139
Bromothymol blue, **I**, 157
   -phenol red, **I**, 139
Brucine sulphate, **I**, 154
   R, 205

Butyl Cellosolve, **So**, 35

CACOTHELINE, **I**, 154
Cadmium, **GD**, cobalticyanide, 61
   2-(*o*-hydroxyphenyl)-benzoxazole,
      110
   tetraphenylmethylarsonium iodide,
      127
   TD, silver nitrate, 205

277

AUTHOR INDEX

283

Date Due